## *Mechanics*



## *Design*



## *Researched Writing*



## *Documentation*



## *Glossaries*

Megan Dudek

Dr. Michael Murphy - See for advice.

# *The Writer's Pocket Handbook*

Alfred Rosa
*University of Vermont*

Paul Eschholz
*University of Vermont*

# *The Writer's Pocket Handbook*

SECOND EDITION

New York San Francisco Boston
London Toronto Sydney Tokyo Singapore Madrid
Mexico City Munich Paris Cape Town Hong Kong Montreal

Senior Vice President and Publisher: Joseph Opiela
Vice President and Publisher: Eben W. Ludlow
Development Editor: Margaret Manos
Marketing Manager: Christopher Bennem
Senior Production Manager: Bob Ginsberg
Project Coordination, Text Design, and Electronic
Page Makeup: Nesbitt Graphics, Inc.
Cover Design Manager: Wendy Ann Fredericks
Cover Designer: Joan O'Connor
Manufacturing Buyer: Lucy Hebard
Printer and Binder: Webcrafters, Inc.
Cover Printer: Coral Graphic Services, Inc.

For permission to use copyrighted material, grateful acknowledgment is made to the copyright holders on pp. 273, which are hereby made part of this copyright page.

Library of Congress Cataloging-in-Publication Data

Rosa, Alfred F.
The writer's pocket handbook/Alfred Rosa, Paul Eschholz.—2nd ed.
p. cm.
Includes index.
ISBN 0-201-78478-5 (pbk.)
1. English language—Rhetoric—Handbooks, manuals, etc. 2. English language—Grammar—Handbooks, manuals, etc. 3. Report writing—Handbooks, manuals, etc. I. Eschholz, Paul A. II. Title

PE1408 .R6755 2002
808'.042—dc21

2002070306



Please visit our website at http://www.ablongman.com

ISBN 0-201-78478-5

12345678910—WC—05040302

# How to Use This Book

Here's how to find what you are looking for in this book:

- **Use the Brief or Detailed Tables of Contents.** A brief table of contents appears inside the front cover and on the facing page. This will lead you to the section that you want. Once in the appropriate section, use the elements of the page to guide you to the specific information you need. On occasion you may wish to use the detailed table of contents at the end of the book and on the inside back cover.
- **Use the index.** This alphabetical listing of the topics, terms, and usage items begins on page 259.
- **Use the directories to documentation models.** Directories for citation and bibliographic models appear in the sections for each documentation style—MLA (pp. 178–196), APA (pp. 203–213), *Chicago Manual* (pp. 220–227), and CSE (pp. 233–237).
- **Use the glossaries.** Use the glossary of usage (pp. 242–249) to help you use certain words and expressions correctly, such as infer and imply. Use the glossary of grammatical terms (pp. 249–257) for definitions of technical language.
- Use the elements of the page.

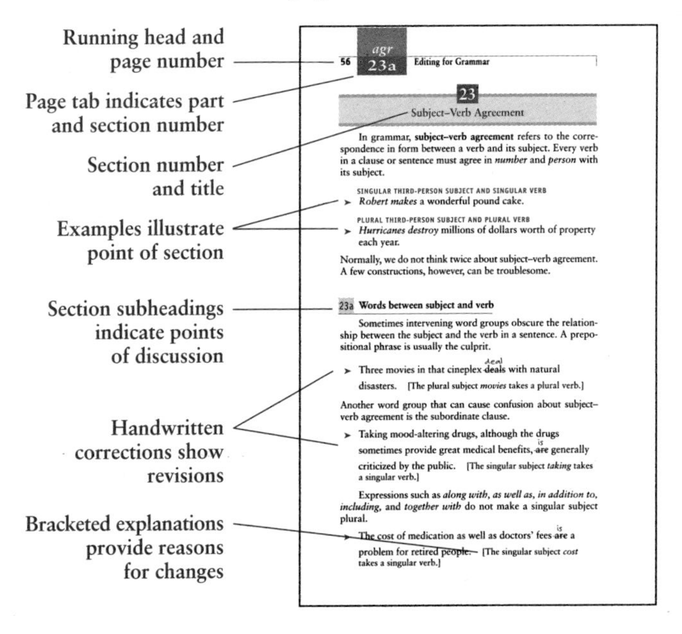


56 agr 23a Editing for Grammar

23 Subject–Verb Agreement

In grammar, **subject–verb agreement** refers to the correspondence in form between a verb and its subject. Every verb in a clause or sentence must agree in *number* and *person* with its subject.

SINGULAR THIRD-PERSON SUBJECT AND SINGULAR VERB

➤ *Robert makes* a wonderful pound cake.

PLURAL THIRD-PERSON SUBJECT AND PLURAL VERB

➤ *Hurricanes destroy* millions of dollars worth of property each year.

Normally, we do not think twice about subject–verb agreement. A few constructions, however, can be troublesome.

**23a Words between subject and verb**

Sometimes intervening word groups obscure the relationship between the subject and the verb in a sentence. A prepositional phrase is usually the culprit.

➤ Three movies in that cineplex ~~deals~~ deal with natural disasters. [The plural subject *movies* takes a plural verb.]

Another word group that can cause confusion about subject–verb agreement is the subordinate clause.

➤ Taking mood-altering drugs, although the drugs sometimes provide great medical benefits, ~~are~~ is generally criticized by the public. [The singular subject *taking* takes a singular verb.]

Expressions such as *along with, as well as, in addition to, including,* and *together with* do not make a singular subject plural.

➤ The cost of medication as well as doctors' fees ~~are~~ is a problem for retired people. [The singular subject *cost* takes a singular verb.]

# *Writing with Purpose*

# 1 Analyzing the Writing Task

## 1a Understanding the writing assignment

Much of your college writing will be in response to very specific assignments. Your environmental studies professor, for example, may ask you to write a paper in which you report both the pro and con arguments for the greenhouse effect; your American history professor may assign a paper asking you to explain the long-term effects of Japanese-American internment during World War II. From the outset you need to understand precisely what your instructor is asking you to do. The keys to understanding assignments such as these (or exam questions, for that matter) are *subject* words (words that focus the content) and *direction* words (words that indicate your purpose for and method of development in writing). In the first example above, the subject words are *greenhouse effect* and *pro and con arguments.* The direction word is *report.* In the second example, the subject words are *long-term effects* and *Japanese-American internment during World War II.* The direction word is *explain.*

While we are familiar with most direction words, we are not always sure what they call for and how they are distinguished from each other. Use the following list of direction words and their meanings to help you analyze paper assignments and exam questions:

*Analyze:* take apart and examine closely
*Argue:* make a case for a particular position
*Categorize:* place into meaningful groups
*Compare:* look for differences; stress similarities
*Contrast:* look for similarities; stress differences
*Critique:* point out positive and negative features
*Define:* provide the meaning for a term or concept
*Describe:* give detailed sensory perceptions of a person, place, or thing
*Evaluate:* judge according to some standard
*Explain:* make plain or comprehensible
*Illustrate:* show through examples

*Interpret:* explain the meaning of something
*List:* catalog or place in sequence
*Outline:* provide abbreviated structure for key elements
*Prove:* demonstrate truth of by logic, fact, or example
*Review:* summarize key points
*Synthesize:* bring together or make connections among elements
*Trace:* delineate a sequence of events

Test your recognition of both subject words and direction words in the following assignments. In your own words, what is the writer being asked to do in each instance?

> Write an essay in which you analyze three scenes from the movie *Casablanca* to illustrate propaganda and/or cultural values of the early 1940s.
>
> Write a paper in which you describe the way recycling is carried out in your dining hall, clearly identifying all the steps involved, and then argue for a way to make the process more effective.
>
> Write a paper defining the terms *ageism* and *sexism.* What does each term mean? Explain the similarities you see in the ways each phenomenon operates in our society.

## 1b Choosing your own paper topic

Sometimes instructors allow you to write on any course-related subject that interests you. You can begin by asking yourself these questions about the subject matter of your course: What do I really care about? What am I interested in? What do I know something about? What do I want to learn more about? Your answers to these questions will provide you with potentially good subjects. Resist the temptation to seize the first subject that comes to mind; take your time and review all the possibilities because selecting a suitable topic may be the most important task in the writing of your paper. A clearly focused topic makes researching and writing a paper manageable and the likelihood of solid—even exciting—results possible. One effective way of focusing on a topic is to brainstorm by asking some specific questions about your subject. At the start of her writing project, one environmental studies student chose recycling paper waste in the United States as her subject and brainstormed the following list of possible questions:

> How serious is our paper waste problem?
>
> How much of our paper is currently being recycled?
>
> Is recycling paper waste an economic or an environmental issue?
>
> Should Americans be required to recycle household paper waste?
>
> What are the problems of collecting waste paper?
>
> What is currently being done with old newspapers and telephone books?
>
> What is made from recycled paper?
>
> Do newspapers use recycled newsprint?
>
> If we don't recycle old newspapers, what will be the impact on our environment?
>
> Should we concentrate on recycling nonbiodegradable wastes first?

Each question narrows the general subject area suggesting a more manageable topic for a research project. Also, simply phrasing your topic as a question gives you a starting point; that is, if you think in terms of a question, your work has direction from the outset because you are focusing on answering your question.

## 2 Generating Ideas and Collecting Information

Ideas and information lie at the heart of good prose. Ideas grow from information (facts and details); information supports ideas. To inform and intellectually stimulate your readers, gather as many ideas and as much specific information as you can about your topic. If you try to write before doing such work, you run the risk of producing a shallow, boring draft.

## 2a Brainstorming

Collect ideas by *brainstorming,* listing the things you already know about your topic, in no particular order. Freely associate one idea with another; let your mind take you in whatever direction it will. Try not to censor yourself or edit your brainstorming because you simply do not know what will emerge or how valuable it might be in the end. Write quickly. Do not worry about spelling or punctuation—abbreviate. Keep your list over several days, adding new thoughts as they come to you. Consider the following brainstorming list.

Animal Rights

animals are living beings, too.
they have nervous systems so they feel pain
questionable experiments—proving the obvious, duplicative
some medical experiments nec., but only when no alternatives are available
computer models instead
also in vitro and membrane tests
Nat. Cancer Inst. reduced its use of animals 95 percent by using in vitro tests
cosmetic companies—Draize eye test
household product testing (lethal dose test for detergents, shampoo, etc.)
fur belongs on animals
steel jaw traps—animals chew off legs
dogs/cats caught, too
progress made—legislation against inhumane traps
some cosmetic companies quit using Draize, consumer protest
puppy mills
spay/neuter—the answer to pet overpopulation

## 2b Asking questions

Ask questions about your topic to discover areas for exploration and development. The newspaper reporter's five *Ws* and an *H*—Who? What? Where? When? Why? and How?—are excellent questions to start with, keeping in mind that questions give rise to yet more questions. Note that every set of questions will vary with the topic and with the person formulating them. Here is one sample set for echinacea, a popular medicinal herb:

1. *Who* uses it?
2. *What* does it do?
3. *Where* is it sold?
4. *When* did it become popular?
5. *Why* do people use it?
6. *How* can I find out if it's safe?

## 2c Clustering

*Clustering* allows you to generate material and to sort it into meaningful groupings. Put your topic, or a key word or phrase about your topic, in the center of a sheet of paper and draw a circle around it. (The example on page 7 shows the topic, gun control, in the center.) Draw four or five (or more) lines radiating out from this circle, and jot down main ideas about your topic; draw circles around them as well. Repeat the process by drawing lines from the secondary circles and adding examples, details, and any questions you have. You may find yourself pursuing one line of thought through many connected circles before beginning a new cluster. Do whatever works for you. As with brainstorming, keep writing—do not stop to think about being neat or punctuating correctly.

## 2d Keeping a journal

Many people find their best ideas when they are not consciously working on a writing assignment. Carry a notebook with you to record thoughts and observations, bits of overheard conversation, ironies, insights, and interesting facts and statistics from newspapers and magazine articles. When it comes time to write, your journal will be a ready resource for ideas, examples, quotations, and references to books, journals, and newspaper articles, and whatever else you might have found useful to enhance your writing.

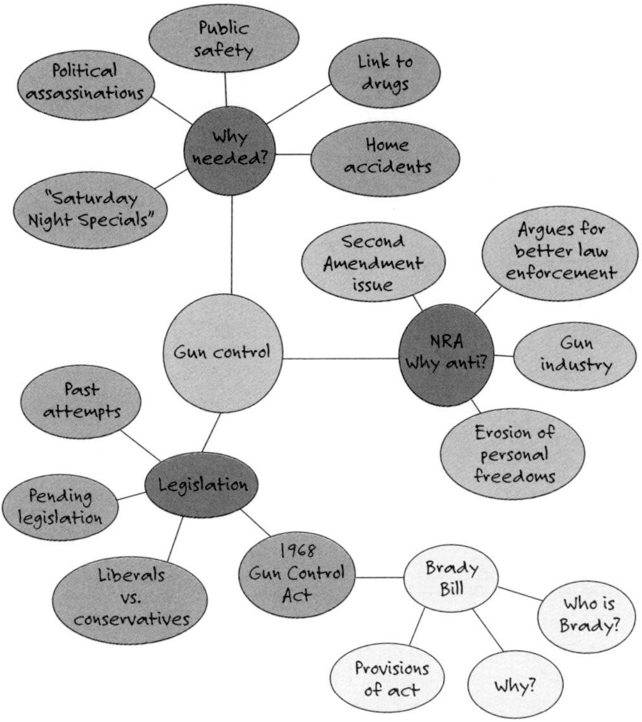

## 2e Freewriting

*Freewriting* is simply writing for a brief, uninterrupted period of time—say, five or ten minutes—on anything that comes into your mind. It is a way of getting your mind working by easing into the writing task. Start with a blank sheet of paper or blank computer screen and write as quickly as you can without stopping for any reason whatsoever. Write as if you were talking to your best friend. If you run dry, don't stop; repeat the last few things you wrote or write, "I have nothing to write," over and over again, and you'll be surprised—writing with more content will begin to emerge. Once you have become comfortable with open-ended freewriting, you can move to more focused freewriting in which you write about a specific aspect of your topic. By freewriting regularly, you come to feel more natural and comfortable about your writing.

## 2f Researching

You may sometimes want to supplement what you know about your topic with research. This does not necessarily mean formal library work. Firsthand observations and interviews with people knowledgeable about your topic are also forms of research and are usually more up to date than library research. Whatever your form of research, take careful notes, so you can accurately paraphrase an author or quote an interviewee.

## 2g Rehearsing ideas

Try rehearsing what you are going to write by running ideas or phrasings, even sentences, through your mind until they are fairly well crafted before you transfer them to paper. The image of the writer, pencil in hand, staring off into space, perhaps best captures the essence of this technique. Rehearsing may suit your personality and the way you think. Moreover, because it is thoughtful practice, rehearsing may help you generate ideas. Sometimes rehearsing may even be done orally. Try taking ten or fifteen minutes to talk your way through your paper with a roommate, friend, or family member.

## 2h Visualizing topics

Some experts believe that a large part of our thinking is *visual*—done through images. Tapping into those images can be a productive way of developing your ideas. For example, if you wish to describe an ancient pueblo, visualizing one you recently visited in New Mexico can make your task easier. Imagining that pueblo can also lead you to yet more images of what life is like there.

## 2i Thinking creatively

There are many definitions of creativity, but creativity, one way or another, involves moving beyond what is generally regarded as normal or expected. To push an idea one step further, to make a connection not easily recognized by others, to step to one side of your topic and see it in a new light, to ask a question no one else would, to arrive at a fresh insight is to be *creative*. Creativity and inspired thinking are within the reach of most writers if they take the writing process seriously and work hard.

# 3 Determining a Purpose

Your purpose is the answer you give the question "What am I trying to accomplish in this composition?" Being clear about your purpose helps you choose the best supporting details and arrange them in the most effective order.

Generally, nonfiction writing has one of four purposes: (1) to make discoveries about oneself and the world, (2) to express the writer's thoughts and feelings about a life experience, (3) to inform readers by explaining something about the world around them, or (4) to persuade readers to some belief or action.

## 3a Writing to discover

In writing to discover you follow language into new territory, into topics and ideas that are new to you. The English novelist E. M. Forster asked, "How do I know what I think until I see what I say?" He understood that writing holds up a mirror both to the world and to you, the writer, almost magically leading you to insights and revelations. Through writing, you now realize what you never knew before. Meaning emerges from ideas that in turn beg for information and examples. Juxtapositions shed light. You now perceive the way things work, or realize how parts make up wholes, or appreciate why people think and act the way they do.

Nowhere are the results of writing to discover more evident than in your first draft, hence the reason it is often called the "discovery draft." Whether writing a letter, a journal entry, an essay, or a full-blown research paper, whether writing a first draft or a revision—take risks, entertain the unfamiliar, get inside your topic, go in a new direction, and allow complexity to evolve and to build. And be prepared for surprises.

## 3b Writing from experience

In writing from experience, or writing expressively, you put your thoughts and feelings before all other concerns. When you express yourself about what it felt like to turn eighteen, describe the relationship you have with your father, narrate a camping experience you had with a friend, or share an insight you had about the career you want to pursue, you are writing from experience. The first purpose of expressive writing is, therefore, to clarify life's experiences, and the second purpose is to communicate what you learn to someone else. Expressive writing is immensely appealing to readers; the reflections of a thoughtful and sensitive writer illuminate the reader's experiences and clarify his or her own feelings and ideas. Here, for example, are the reflections of a writer on her ambitious nature:

> ➤ I've always liked ambitious people, and many of my closest friends have had grandiose dreams. I like such people, not because I am desperate to be buddies with a future secretary of state but because I find ambitious people entertaining, interesting to talk to, and fun to watch. And, of course, I like such people because I am ambitious myself, and I would rather not feel apologetic about it. —Perri Klass, "Ambition"

## 3c Writing to inform

Informative writing focuses on the world outside the writer—the events, people, places, things, and ideas in the *objective* or *real world.* In informative writing you report, explain, analyze, define, classify, compare, describe a process, or get at causes and effects. Informative writing is the kind most often found in newspaper and magazine articles and nonfiction books.

The following example of informative writing provides useful information about the South:

> ➤ To the visitor the South looks uniform and cohesive, but this is misleading. There are many Souths. The Englishness of colonial Virginia plantations is very different from the Spanish flavor of Florida's St. Augustine or the French bouquet of the bayou region of Louisiana.
>
> —Parke Rouse, Jr., "The Old South Way of Life: America's Most Colorful Region"

## 3d Writing to persuade

In writing to persuade, you attempt to influence your reader's thinking and attitudes toward a subject or issue and sometimes move him or her to a particular course of action. Persuasive writing uses logical reasoning, authoritative evidence and testimony, and sometimes emotionally charged language and examples.

> ➤ Physicians and nurses often assume that pain is not a problem unless the patient or family brings it up. If the patient is stoical and trying not to be a "bother," then the pain and unnecessary suffering continue. [. . .]
>
> Not wanting to "bug" the healthcare team is a natural reaction. However, the control of pain contributes to a person's physical and emotional well-being. And that is the goal toward which the healthcare team, the patient, and the family are all working.
>
> —Alexandra Beebe, "The Family: Important Participants in Cancer Pain Control"

Most of the writing that you do in college will be informative and some will be persuasive in character; occasionally you may be asked to write from experience. Often you will use some combination of these types of writing in a single composition.

## 4 Establishing a Thesis Statement

Once you have generated ideas and information about your topic by asking questions, brainstorming, and researching, you are ready to begin organizing your thoughts. At this stage, you must commit to a controlling idea, a thesis. The thesis of a prose composition is its main idea, the point it is trying to make. The thesis is often expressed in one or two sentences called a *thesis statement.*

➤ The whole language approach to teaching reading best serves the needs of all children by equipping them with a well-rounded understanding of language and a variety of strategies for decoding it. —Michelle Whalen, student

The thesis statement should not be confused with a purpose statement. Whereas a thesis statement makes an assertion about your topic, a purpose statement describes what you are trying to do in the paper. Normally, it is not a good idea to include a statement of your purpose in the paper itself.

➤ I plan to explain why the whole language approach to teaching children to read is better than the phonics approach.

A thesis statement should be

1. more general than the ideas and facts used to support it,
2. appropriately focused for the length of your paper, and
3. the most important point you make about your topic.

A thesis statement should not be a question but an assertion—a claim made about a debatable issue that can be supported with evidence.

An effective strategy for developing a thesis statement is to begin by writing *What I want to say is that. . . .*

➤ *What I want to say is that* unless language barriers between patients and health care providers are bridged, many patients' lives in our nation's most culturally diverse cities will be endangered.

Later, you can delete the formulaic opening, and you will be left with a thesis statement.

➤ Unless language barriers between patients and health care providers are bridged, many patients' lives in our nation's most culturally diverse cities will be endangered.

The thesis statement is usually set forth near the beginning of the composition, sometimes after a few sentences that establish a context.

If you find yourself writing a question for a thesis statement, answer the question first and then write your statement. The answer to your question often provides you with a preliminary thesis statement, a one- or two-sentence rendering of your main argument. The student who generated the questions listed earlier kept returning to the question of what Americans do with their old newspapers and telephone books. Her initial answer to that question—her preliminary thesis statement—directed the course of her research in the library:

➤ Americans are currently collecting more old newsprint than they can recycle.

As she analyzed the information she found, she shifted her focus to the uses of old newsprint. This shift, in turn, caused her to reformulate her thesis statement. In the end, she decided to take a much stronger, more argumentative position.

➤ To solve the tremendous newspaper waste problem, Americans must find long-term uses for old newsprint that are environmentally and economically sustainable.

## 5 Analyzing Your Audience

Having decided on a purpose for writing and a thesis statement, it is time to consider your audience or intended readership. Students often mistakenly assume their instructor is their only audience. Though it is true that your teacher will read your composition, do not forget the other students in your class. They, after all, make up the writing community to which you belong.

Use the following list of questions to identify your audience so that you can make appropriate decisions on content, sentence structure, and word choice.

**AUDIENCE QUESTIONS**

- Who are my readers?
- Is my audience specialized (my chemistry lab partners, other Macintosh computer users) or general (literate adults)?
- What do I know about my audience (age; sex; amount of education; religious, social, economic, and political attitudes)?
- What does my audience know about my subject? What is its knowledge level—expert or novice?
- What does my audience need to know that I can tell it?
- Will my audience have misconceptions that I can clarify?
- What is my relationship to my audience: Friendly? Objective? Hostile?
- How will my audience respond to what I have to say (interested, open-minded, resistant, hostile)?
- Is there any specialized language my audience needs or that I should avoid?
- What do I want my audience to do? How can I help my audience?
- How should I sound—formal or informal?

The best writers know their readers. They try to see issues as their readers might and construct their arguments accordingly.

## 6 Crafting Effective Paragraphs

The paragraph has a topic sentence that states or implies the main idea of the paragraph. The well-written paragraph is *unified*—all sentences relate to the main idea; it is well *developed*—there is representative and sufficient supporting detail; and it is *coherent*—ideas and sentences flow logically and smoothly.

The following paragraph on the life cycle of neighborhoods is an example of a well-written paragraph.

➤ Neighborhoods are often assigned human characteristics, one of which is a life cycle: they have a birth, a youth, a middle age, and an old age. A neighborhood is built and settled by young, vibrant people, proud of their sturdy new homes. Together, residents and houses mature as families grow larger and additions get built on. Eventually, though, the neighborhood begins to show its age. Buildings sag a little, houses aren't repaired as quickly, and maintenance slips. The neighborhood may grow poorer as the young and upwardly mobile find new jobs and move away, while the older and less successful inhabitants remain. —Kevin Cunningham, student

Cunningham provides a clear topic sentence that explains what he means by a life cycle. He then goes on to describe the stages of that cycle and the process by which one stage leads to the next. Cunningham maintains unity by excluding anything that does not directly relate to the concept of the life cycle. He achieves coherence by ordering his sentences to flow smoothly without breaks in rhythm or meaning and by using transitional words (*and, as, eventually, while*).

## 6a Unity

In a unified paragraph, all sentences relate to the topic sentence by explaining it with supporting details. You should eliminate all sentences that stray from the point of the topic sentence. In the following paragraph, the sentences that violate paragraph unity are italicized.

NON-UNIFIED

➤ When I was growing up, one of the places I enjoyed most was the cherry tree in the back yard. *Behind the yard was an alley and then more houses.* Every summer when the cherries began to ripen, I used to spend hours high in the tree, picking and eating the sweet, sun-warmed cherries. *My mother always worried about my falling out of the tree, but I never did.* But I had some competition for the cherries—flocks of birds that enjoyed them as much as I did and would perch all over the tree, devouring the fruit whenever I wasn't there. I used to wonder why the grown-ups never ate any of the cherries; but actually, when the birds and I had finished, there weren't many left. —Nancy Shafran, student

In the following paragraph, the topic sentence comes first, and the anecdote that follows illustrates this topic sentence, thus unifying the paragraph.

**UNIFIED**

➤ Bees are filled with astonishments, confounding anyone who studies them, producing volumes of anecdotes. A lady of our acquaintance visited her sister, who raised honeybees in northern California. They left their car on a side road, suited up in protective gear, and walked across the fields to have a look at the hives. For reasons unknown, the bees were in a furious mood that afternoon, attacking in platoons, settling on them from all sides. Let us walk away slowly, advised the beekeeper sister, they'll give it up sooner or later. They walked until bee-free, then circled the fields and went back to the car, and found bees there, waiting for them.

—Lewis Thomas, "Clever Animals"

## 6b Development

Whether it is first or last or someplace in the middle, whether it is explicitly stated or merely implied, the topic sentence is the point of departure for writing an effective paragraph. How you develop—that is, how you clarify and support through examples and explanations—your topic sentence depends on how you answer the question "Why or how is this so?" Consider, for example, the following topic sentence:

➤ Neat people are especially vicious with mail.

Applying the question "Why or how is this so?" suggests a strategy for development: the writer needs to give us examples of how "neat people are vicious with mail."

➤ Neat people are especially vicious with mail. They never go through their mail unless they are standing directly over a trash can. If the trash can is beside the mailbox, even better. All ads, catalogs, pleas for charitable contributions, church bulletins, and money-saving coupons go straight into the trash can without being opened. All letters from home, postcards from Europe, bills and paychecks are opened, immediately responded to, then dropped in the trash can. Neat people keep their receipts only for tax purposes. That's it. No sentimental salvaging of birthday cards or the last letter a dying relative ever wrote. Into the trash it goes.

—Suzanne Britt, "Neat People vs. Sloppy People"

How long should a paragraph be? In expository prose (prose that explains), a paragraph is typically 100 to 150 words. Longer paragraphs appear in professional journals that discuss new or highly complex information requiring in-depth explanation and evidence. Paragraphs are shorter in newspapers and magazines where the print column is narrow and needs to be broken more often and where the subject matter is less demanding of the reader. Occasionally writers use one- or two-sentence paragraphs to transition from one part of an essay to another, to make a dramatic point, or to highlight the content of the beginning or ending.

Sometimes a paragraph is too long and needs to be divided or restructured for clarity. The more common problem is the underdeveloped paragraph that leaves the reader wanting more information. To be sure you have provided enough information to enlighten and convince the reader—and thus communicate all you intend—ask a friend to read your composition and point out anything that is not explained clearly or fully enough.

Often logic dictates that you use a combination of methods to develop your topic sentence. In the paragraph that follows, the writer uses cause and effect, informational process analysis, and a hypothetical case in point (a kind of example) to develop her topic sentence, *They [cockroaches] reproduce at a truly amazing rate.*

➤ Cockroaches give credence to the old adage that there is safety in numbers. They reproduce at a truly amazing rate. About two months after mating, a new generation of cockroaches is born. One cockroach can produce about two dozen offspring each time it mates. To get some idea of their reproductive power, imagine that you start with three pairs of cockroaches that mate. Approximately three weeks after mating the females lay their eggs, which hatch some forty-five days later. If we assume two dozen eggs from each female, the first generation would number seventy-two offspring. These roaches would continue to multiply geometrically so that by year's end the colony's population would total more than ten thousand cockroaches. Stopping the process is almost impossible because even if we annihilate the adult population, a new generation still in egg form would be in the making. —Courtney Smith, student

## 6c Coherence

Coherence in a paragraph is achieved by arranging sentences in the most effective order, using transitional words and phrases, repeating key words and phrases, and using parallel structures. These techniques knit the sentences of a paragraph into a tight structure. As originally written, the following paragraph lacked coherence; it seemed to jump from idea to idea. In the revision, coherence is achieved by (1) repositioning a sentence, (2) repeating a key term, and (3) adding a transition.

➤ In comparison with other parts of the world—Mexico, India, and Africa, for example—the United States does not have widespread poverty. That we are the richest and most technologically advanced of all nations, however, makes the poverty that exists seem all the more surprising and incongruous. We have poor people among the elderly, minorities, mentally ill, and children. As the great historian Arnold Toynbee so succinctly said, "The worst country to be poor in is America." Some of our poor people have clothing, appliances, radios, televisions, furniture, and housing that would make the poor of other countries envious. But ~~We~~ we also have poor people who are homeless (including runaway teenagers) and ill nourished.

### Transitional words and phrases

Transitional words and phrases connect units of thought—within a sentence, between sentences, and between paragraphs. In other words, transitions signal relationships and thus are categorizable by their functions. The following list presents some of the more common transitional words and expressions categorized by meaning (some words appear in more than one category because their meaning varies with context):

ADDITION again, also, and, besides, further, furthermore, in addition, moreover, too

CAUSE AND EFFECT accordingly, as a result, consequently, hence, so, then, therefore, thus

COMPARISON by comparison, likewise, similarly

CONCESSION although, although this may be true, certainly, even though, granted, it is true, naturally, of course, to be sure, to tell the truth, with the exception of

CONTRAST after all, at the same time, but, conversely, however, in contrast, in spite of, nevertheless, on the contrary, on the other hand, still, yet

EXAMPLE for example, for instance, one case of

TIME afterward, at the same time, currently, earlier, immediately, in the meantime, lately, later, shortly, subsequently, this time, until now

PLACE above, around, below, beyond, elsewhere, farther on, here, nearby, opposite, there

CLARIFICATION actually, in fact, in other words, in simpler terms, partly, simply stated, that is, to put it differently

SEQUENCE first, second, third [*not* firstly, secondly, and so on], finally, following, in time, next, then

SUMMARY in brief, in conclusion, in short, to conclude, to summarize

The transitional words and phrases in the following paragraph are italicized. To test how important they are in showing relationships among the ideas in the paragraph, try omitting them or substituting others.

➤ In America, *of course,* social distinctions have never been so sharp as they are in England. We find it somewhat easier to rise in the world, to move into social environments unknown to our parents. This is possible, *partly,* because speech differences are slighter; *conversely,* speech differences are slighter because this is possible. *But* speech differences do exist. If you've spent all your life driving a cab in Philly and, having inherited a fortune, move to San Francisco's Nob Hill, you will find that your language is different, perhaps embarrassingly so, from that of your new acquaintances.

—Paul Roberts, "Speech Communities"

# 7 Revising

When you have finished writing your draft, give it an honest appraisal. Focus on the large issues—thesis, purpose, content, organization, and paragraph structure—that affect your entire composition. It would be counterproductive to look at grammar and punctuation, for example, if the elements that make an essay "go" need work. So it is with writing. First you revise—work on the large issues that clarify your purpose and improve your organization—and then you edit—check for correctness and style.

## 7a Revising the largest elements

Revision is best done by asking yourself questions about what you have written. Otherwise, you can stare at a draft for a good long time, wondering what you should be looking for. Begin by reading, preferably aloud, what you have written, paying attention to every single word, and looking for lapses in the logical flow of thought. Resist the temptation to plunge immediately into a second draft. Try to look at your writing as a whole and to tackle your writing problems systematically.

Even though you are the writer, step back from the paper and critique it as a reader might. Use the questions in the Checklist for Revising Large Elements to assess whether your meaning is clear, and whether you have effectively communicated it to the reader.

**Checklist for Revising Large Elements**

1. Is the topic well focused? State the topic in a single sentence.
2. Is the purpose to express, inform, or persuade?
3. Underline the thesis statement. Does it clearly identify the topic and make an assertion about it?
4. Is the organizational pattern the best one given the purpose? What alternative pattern would be more effective?
5. Are there enough supporting details, and are the examples well chosen to support the thesis?
6. Does the paper achieve its purpose?
7. How effective is the beginning? Ending?
8. Is the title the best one for this paper?

In answering these questions, you may discover that parts of your paper bear little or no relationship to your thesis and purpose. You may need to rearrange your examples for greater impact. Or perhaps you need a transition between paragraphs.

## 7b Revising your sentences and diction

Having revised the largest elements of your composition, turn next to your sentences and your diction. The best approach to considering these elements is to look at your sentences one at a time, again asking yourself questions from a reader's perspective. Use the Checklist for Revising Sentences and Diction.

**Checklist for Revising Sentences and Diction**

1. Does each sentence convey its thought clearly?
2. Does the writer subordinate less important ideas to more important ones?
3. Is each sentence a complete sentence, or has the writer written any unintentional sentence fragments?
4. Is any sentence a comma splice or run-on?

(*continued*)

Checklist (*continued*)

5. Does the writer use a variety of sentence structures?
6. Is the diction exact, with each word meaning precisely what it should?
7. Does the writer engage the reader with concrete nouns and strong action verbs?
8. Does the writer avoid wordiness and clichés?
9. Has the writer committed any usage errors? See the Glossary of Usage.

You may find that some of your sentences are long and rambling and that others are short and choppy, giving the impression that your thoughts are disconnected. Perhaps you shifted focus within some sentences or used the same sentence pattern throughout most of your composition. Sentence problems such as these may drive you to reconsider paragraphs you previously thought were effective.

# *Writing with Clarity*

# 8 Strive for Conciseness

Conciseness in writing is not necessarily a function of the length of your sentences. A long sentence can be concise and a short sentence wordy. Wordiness occurs when you have words that contribute little to the meaning of your sentence and at the same time overshadow those parts of the sentence that carry the meaning. Concise writing lets your meaning shine through regardless of the length of your sentences.

## 8a Subjects and verbs

The essence of a sentence is its subject and its verb. The *subject* identifies who or what the sentence is about, and the *verb* indicates the subject's action or state of being. Sentences often lose their conciseness because the actor that should be identified in the subject and the action that should be conveyed by the verb are lost in weak language or buried in other parts of the sentence.

➤ The *aim* of college athletics *is* the promotion of individual achievement, teamwork, and school spirit. [*Aim* and *is* are weak.]

➤ College *athletics promotes* individual achievement, teamwork, and school spirit. [*Athletics* and *promotes* are specific and vivid.]

### Strong verbs

Strong verbs energize writing by vividly picturing the action in each of your sentences. When you encounter one of these weak verbs while revising, look for a stronger verb.

| | |
|---|---|
| is, are, was, were, will be | add |
| have, has, had | involve |
| deal with | concern |
| make | reflect |
| give | provide |
| do | become |
| use | go |
| get | |

### Active voice

The *active voice* emphasizes the actor in a sentence; the *passive voice,* the receiver of the action.

tornado flattened the

➤ The grain elevators ~~were flattened by the tornado.~~

When the action itself is what is important or you wish to minimize the importance of the actor, use the passive voice. The passive voice is often used in scientific and technical writing to emphasize processes and events rather than individuals.

➤ Three drops of boric acid *were added* to the solution.

➤ Investigative teams *were sent* by the Centers for Disease Control.

### *There is, it is*

Although necessary in references to time and weather, expletive constructions clutter your sentences with unnecessary words.

Many followed

➤ ~~There were many~~ peaceful demonstrations ~~following~~ the execution.

Note that when you eliminate an expletive, you also give your sentence a specific subject and an action verb.

### Combined sentences

Look for opportunities to combine the contents of two or more related sentences into one concise sentence.

WORDY

➤ Hector hopes to get many hours of overtime work. He assembles cars at the local General Motors plant. The General Motors plant has just been awarded two large government contracts.

CONCISE

➤ Hector hopes to get many hours of overtime work assembling cars at the local General Motors plant, which has just been awarded two large government contracts.

## 8b Redundancies

Too often, unnecessary repetition creeps into writing and distracts the reader. Have you ever written *purple in color, decided in my mind,* or *end result*? Edit by deleting words.

➤ We reviewed the basic ~~and fundamental~~ principles of public speaking.

➤ The supermarket ~~where people buy groceries~~ extended its hours.

## 8c Empty words and phrases

In strong writing, every word carries its own weight. Cut empty words or phrases that add little or nothing to your meaning. Omit them entirely, or use a more precise word.

➤ ~~The area of l~~Linguistics attracts researchers from many disciplines.

➤ ~~In my opinion, the~~ The orchestra performed ~~really~~ well.

Two of the worst culprits are vague nouns (*area, aspect, factor, situation,* and *type*) and vague modifiers (*basically, major, really, very,* and *virtually*). Look for these other empty words and phrases as you edit:

apparently, seemingly
essentially
for all intents and purposes
generally
in some ways
I think, I feel, I believe
it seems to me
kind of, sort of
tend to
various

## 8d Inflated expressions

Wanting to sound serious or knowledgeable, writers sometimes use expressions such as *at this point in time* (instead of *now*) or *in the event that* (instead of *if*). Try to reduce these expressions to a single, more precise word.

because

➤ We were late ~~on account of the fact that~~ the bus broke down.

influenced

➤ Professor Trent ~~had an effect upon~~ my decision to major in anthropology.

Other inflated expressions that can be reduced to a word or two include the following:

| INFLATED | CONCISE |
|---|---|
| by means of | by |
| due to the fact that | because |
| for the purpose of | for |
| for the simple reason that | because |
| in order to | to |
| in spite of the fact that | even though, although |
| in today's society | today, now |
| it is important that, it is necessary that | must |
| on the occasion of | when |
| prior to, in anticipation of | before |
| until such time as | until |
| with regard to | about |

## 8e Other clutter

You can often tighten your prose by converting clauses to phrases.

➤ Orange juice ~~that is~~ made from concentrate is not as good as fresh squeezed.

➤ Spin doctors, ~~who are~~ people who manipulate the news media, appeared after the governor gave her speech.

Sometimes a clause or phrase can be converted to a single word.

noisy

➤ The spectators~~, who were noisy,~~ booed the referee's call.

# 9 Balance Parallel Ideas

Use parallel grammatical form to reinforce parallel ideas.

BALANCE A WORD WITH A WORD

➤ I see one-third of the nation *ill-housed, ill-clad,* and *ill-nourished.* —Franklin D. Roosevelt

BALANCE A PHRASE WITH A PHRASE

➤ It's much easier to write *a solemn book* than *a funny book.* —Fran Lebowitz

BALANCE A CLAUSE WITH A CLAUSE

➤ *I never write "metropolis" for seven cents because I can get the same price for "city"; I never write "policeman" because I can get the same money for "cop."* —Mark Twain

## 9a Parallel constructions with *and, but, or, nor, yet*

➤ The lawyer chose not to criticize the judge but ~~appealed~~ to appeal the decision.

➤ A seeing-eye dog announces danger, insures mobility, and ~~is a companion.~~ provides companionship.

To emphasize parallel structures, repeat an article (*a, an, the*), a preposition, the sign of the infinitive (*to*), or the first word of a long phrase or clause.

➤ Timothy wanted a promotion and a salary increase.

➤ Kim had real doubts—to go or to stay.

## 9b Parallel constructions with *either/or, neither/nor, not only/but also, both/and,* and so on

➤ He was neither a good host nor ~~someone who could tell a good story.~~ a good storyteller.

➤ The couple behind us who kept whispering were both inattentive and rude. ~~not showing good manners.~~

## 9c Parallel constructions with *than* or *as*

➤ I would rather resign than agree ~~agreeing~~ to an unethical solution.

Seeing is as good as believing. ~~to believe.~~

# 10 Eliminate Confusing Shifts

Abrupt or inappropriate shifts—changes from first person to third person, from past to present tense, or from informal to formal diction, for example—weaken the logic of your writing or obscure your meaning.

## 10a Shifts in person and number

Shifts in person most commonly occur between first-person pronouns (*I, we, me, us, my, our*) and second-person pronouns (*you, your, yours*) and between second-person and third-person pronouns (*he, she, it, they, him, her, them, his, hers, its, their, theirs*).

➤ We were told we ~~you~~ could ride pack burros to the bottom of the canyon.

➤ If you eat sensibly and watch your fat intake, ~~most people~~ you should be able to maintain your ~~their~~ desired weight.

Most shifts in number—the singular and plural forms of nouns and pronouns—occur when a plural pronoun is used to refer to a singular noun or vice versa.

➤ When students take ~~a student takes~~ an exam, they should read all the questions carefully.

➤ The Organization of American States has not achieved economic unity in the western hemisphere; ~~they have~~ it has succeeded with social and political issues, however.

## 10b Shifts in verb tense

The verb tense in a sentence indicates when the action is taking place. Avoid shifting from one tense to another without a logical reason.

➤ Margaret always stayed at the Algonquin Hotel; she ~~finds~~ found the rooms spacious and beautifully decorated.

## 10c Shifts in mood

Verbs in English have three moods: the *indicative,* the *imperative,* and the *subjunctive.* Problems with inconsistency usually occur with the imperative mood.

➤ In learning to hit a golf ball, take a full back swing, and ~~it is also important to~~ keep your eye on the ball.

## 10d Shifts in subject and voice

Shifting the subject of a sentence (see page 56) and shifting its voice (see pages 67–68) often go hand in hand, resulting in awkwardness and confusion.

➤ We could see the widespread effects of the drought as we approached the farm ~~was approached.~~.

## 10e Shifts from direct to indirect quotation

In direct quotation, the writer gives a speaker's words verbatim—in quotation marks (*The aerobics instructor said, "Let's really raise those knees!"*). In indirect quotation, or reported speech, the writer paraphrases or summarizes what the speaker said (*The aerobics instructor said we should raise our knees higher*). A shift from direct to indirect quotation or vice versa can be confusing.

➤ Our instructor said workstations had to be cleaned at the end of each lab session and ~~report~~ all test-tube breakage^had to be reported.

This example can also be revised to maintain consistent direct quotation.

Our instructor said, "Clean your workstations at the end of each lab session, and report all test-tube breakage."

### 10f Shifts in tone and style

Unnecessary shifts in tone—the attitude or stance the writer takes toward the subject and audience—and in style—the writer's manner of expression—call attention to themselves and take away from the intended meaning. For example, a reader would be surprised by an unexpected shift from a reverent to a sarcastic tone or from a flowery to a simple style. In most cases, such inconsistencies are matters of inappropriate word choice (see pages 37–39).

➤ That some large industrial plants are not conforming to clean-air laws ~~ticks off~~ angers many people.

➤ John and Stefano climbed into the leaky old rowboat and rowed out to the middle of the lake to^fish. ~~reenact the ancient rivalry in which man and fish do battle.~~

## 11 Fix Misplaced and Dangling Modifiers

You can usually move around the words, phrases, and clauses that function as modifiers in a sentence, so place them carefully to avoid unintentionally confusing—or amusing—your reader.

Because most English sentences depend heavily on word order for meaning, place modifiers as close as possible to the words they modify.

## 11a Misplaced modifiers

### Limiting modifiers

Place adverbs such as *just, only, almost, even, hardly, nearly,* and *merely* directly before the words they modify.

➤ He nearly missed his appointment. ~~nearly.~~

### Phrases and clauses

Place phrases and clauses near the words they modify.

➤ Dana wore an old shirt with holes in it for painting. ~~with holes in it.~~

➤ Mary left in a hurry for a vacation. ~~in a hurry.~~

➤ The waiter brought a steak covered with mushrooms to the young man. ~~covered with mushrooms.~~

### Squinting modifiers

A squinting modifier looks in two directions; that is, it appears to modify both the word it follows and the word it precedes.

➤ The passenger who was badly hurt ~~badly~~ needed help.

Placed after *hurt, badly* modifies both *hurt* and *needed.* When *badly* is placed before *hurt,* the sentence suggests several passengers were hurt and one seriously.

➤ The passenger who was hurt ~~badly~~ needed help badly.

Moving *badly* to modify *needed* communicates that only one passenger was hurt but seriously.

### Split infinitives

A split infinitive occurs when a modifier is placed between the *to* and the verb: *to suddenly realize, to slowly climb.* As a general rule, do not split an infinitive.

➤ The president instructed his advisory board to ~~very carefully~~ analyze the new report very carefully.

If repositioning the offending modifier creates an awkward sentence, recast the original sentence.

ORIGINAL

➤ The Red Cross hopes to greatly increase its number of blood donors this year.

RECAST

➤ The Red Cross hopes to attract many more blood donors this year.

### 11b Dangling modifiers

A dangling modifier is a phrase that does not logically relate to the main part of the sentence. A dangling modifier usually appears at the beginning of a sentence. To eliminate a dangling modifier, name the actor immediately after the dangling phrase.

After sitting down, ^the students began^ the exam ~~began.~~ ^.^

## 12 Strive for Sentence Variety

A series of sentences that are all the same length and that follow the same basic pattern is boring. Sentence variety adds interest to your writing.

### 12a Combining short simple sentences

The following passages both describe the city of Vancouver. Although their content is essentially the same, the first is monotonous because it is made up of a series of simple sentences of nearly the same length; the second is interesting because it is made up of sentences of varying structure and length.

CHOPPY AND MONOTONOUS

➤ Water surrounds Vancouver on all sides. The snow-crowned Coast Mountains ring the city on the northeast. Vancouver has a floating quality of natural loveliness. There is a curved beach at English Bay. This beach is in the shape of a half-moon. Residential high rises stand behind the beach. They are pale tones of beige, blue, and ice-cream pink. Turn-of-the-century houses of painted wood frown upward at the glitter of the office towers.

Any urban glare is softened by folds of green lawns, flowers, fountains, and trees. Such landscaping appears to be unplanned. It links Vancouver to its ultimate treasure of greenness. The treasure is thousand-acre Stanley Park.

**INTERESTING**

➤ Surrounded by water on three sides and ringed to the northeast by the snow-covered Coast Mountains, Vancouver has a floating quality of natural loveliness. At English Bay, the half-moon curve of the beach is backed by high rises in pale tones of beige, blue, and ice-cream pink. Turn-of-the-century houses of painted wood frown upward at the glitter of office towers. Yet any urban glare is quickly softened by folds of green lawns, flowers, fountains, and trees that in a seemingly unplanned fashion link Vancouver to its ultimate treasure of greenness—thousand-acre Stanley Park.

—Veronica Thomas, "Vancouver"

Use any of the following structures to combine your short sentences and to make your writing more interesting.

### Subordinating and coordinating conjunctions

➤ After touring ~~Robert went on a tour of~~ the hospital ~~, He saw~~ and seeing many sick people~~,~~ , Robert ~~He~~ could appreciate the work doctors do ~~, He could appreciate~~ and his own good health~~,~~ , which he ~~He~~ had taken ~~his good health~~ for granted.

### Effective modifiers

Instead of writing separate descriptive statements, combine modifiers in a single graphic sentence.

➤ Martha bought the old, dilapidated house. ~~It was old. It was dilapidated.~~

### Semicolon

This novel is not one of her better works~~, It~~ ; it is ill conceived and badly written.

### Absolute constructions

➤ Trucks ~~crammed~~ cramming us on both sides~~.~~, ~~We~~ we sat in traffic for forty minutes.

CAUTION Do not overuse compound sentences. If you find that you overuse *and*, try the following three methods to bring important ideas into focus and make it easier for the reader to follow your thought.

### Changing a compound sentence into a simple sentence with a modifier

➤ My brother, ~~is~~ a brilliant heart surgeon, ~~and he is brilliant, and he~~ is on the staff of Massachusetts General Hospital.

### Changing a compound sentence into a simple sentence with a compound predicate

➤ Tony grabbed his tennis racket~~,~~ and ~~he~~ jumped into the car.

### Changing a compound sentence into a complex sentence

➤ ~~My~~ Because my computer was broken, ~~and~~ I had to write my report in longhand.

## 12b Varying your sentence openings

Most sentences in English begin with the subject. Consider the different ways in which the following sentence can be rewritten to vary its beginning and add interest.

➤ Passengers partied in the ballroom and did not know that the ship had hit an iceberg.

Begin with a participle, a gerund, or an infinitive.

➤ *Partying* in the ballroom, the passengers did not know that the ship had hit an iceberg.

Begin with a prepositional phrase.

➤ *In the ballroom,* the passengers partied, not knowing that the ship had hit an iceberg.

Begin with a subordinate clause.

➤ *As they partied in the ballroom,* the passengers did not know that the ship had hit an iceberg.

# 13 Use Exact Words

Mark Twain wrote, "The difference between the almost right word and the right word is really a large matter—'tis the difference between the lightning-bug and the lightning." It is not enough merely to come close to saying what you want to say; imprecise word choice, or diction, not only fails to express your intended meaning but also runs the risk of confusing your reader.

## 13a Choosing specific and concrete words

Words can be classified as relatively general or specific, abstract or concrete. General words name groups or classes of objects, qualities, or actions. Specific words name individual objects, qualities, or actions within a class or group. For example, *dessert* is a class of food, and *pie* is more specific than *dessert* but more general than *pecan pie.*

Abstract words refer to ideas, concepts, qualities, and conditions—*love, anger, beauty, youth, wisdom, honesty,* and *sincerity,* for example. Concrete words name things you can see, hear, taste, touch, or smell. *Pancake, rocking chair, sailboat, sandpaper, smog, cow,* and *soda* are all concrete words.

General and abstract words fail to create in the reader's mind the vivid response that concrete, specific words do. Always question the words you choose. If, for example, you are describing a course, is it enough to say the course was *wonderful*? How wonderful? Why was it wonderful? In what ways was it wonderful?

## 13b Replacing clichés with fresh language

Expressions that are predictable and dull are termed *clichés* (trite expressions). If an expression comes too quickly to mind and sounds as if you've heard it before, the chances are you have. Replace clichés and overworked expressions with fresh language. How many of the following clichés have you heard lately?

arrived at the scene
at long last
believe it or not
better late than never
cool, calm, and collected
crystal clear
few and far between
first and foremost
for all intents and purposes
in no uncertain terms
in one fell swoop
it stands to reason
last but not least
proves conclusively
rear its ugly head
rude awakening
sadder but wiser
sneaking suspicion
step in the right direction
the bottom line

# 14 Use Appropriate Words

In choosing how to say something, use words in keeping with your subject, and be sensitive to the needs and feelings of your audience. In other words, strive for an appropriate tone. Ask yourself, "Does my choice of words support the seriousness of my subject?" "Do I make my subject sound more serious than it really is?" "Do I keep my audience at too great a distance from me, or do I make my readers uncomfortable because I try to get too close to them?"

## 14a Choosing an appropriate degree of formality

In informal writing—an e-mail message or short memo, for example—the tone may be informal and reflect the way that people who know one another talk to each other. Informal writing is characterized by simple sentences, contractions (*can't, doesn't*), and conversational words (*get around to it* rather than *do; show up* rather than *arrive*).

In formal writing—a letter of application, a class assignment, or a research report—the tone you adopt and the words you choose to create that tone should be serious and not casual. Formal writing does not use contractions and often employs more complex sentences and a more sophisticated—but not stilted or old-fashioned—vocabulary.

## 14b Writing standard English

Standard English, both spoken and written, is the English used by educators, civic leaders, and professionals in all fields and is the language of the media.

Nonstandard English differs from standard English in its use of fairly simple sentence structures; double negatives (*can't see no* for *can't see any*); unusual vocabulary (*spigot* for *faucet*); and unconventional spelling (*nite* for *night*), pronunciation (*yourn* for *yours*), and grammar (*I be going* for *I am going*). Nonstandard English is as acceptable and functional as standard English if it is used within the appropriate social and regional contexts. It is not acceptable where standard English is required.

Slang is the unconventional, informal language of particular subgroups—street gangs, soldiers, or college students, for example; it is not generally known to all members of society. Moreover, slang terms quickly come into and go out of existence, making communication among groups even more difficult. For example, the term *cat's pajamas,* meaning "something special," as in *He thought she was the cat's pajamas,* is no longer used and not readily understood. Slang words such as *diss, NIMBY,* and *dork* need to be revised.

➤ Linda was ~~bummed~~ upset when she got her grades.

Regional expressions are common to a specific geographic area of the country. For example, in New England, the expression *Let's see how this sugars off* means "Let's see how this turns out." The expression derives from the process of boiling down maple sap to make maple syrup, a traditional activity in the Northeast. Like slang, however, regionalisms, though colorful, are not widely known and can leave your reader wondering about your exact meaning, especially when the sentence provides no context.

➤ He sipped a cold ~~pop~~ soft drink.

## 14c Avoiding technical language

Every trade and profession has its own special vocabulary. For example, all printers know the meanings of the terms *flush left, justify,* and *extra leading,* which all refer to the position of type on a printed page. Technical terms allow professionals to communicate quickly and accurately among themselves. When technical language is used to communicate with a general audience, however, it is considered jargon and is inappropriate. It is far better for a doctor in communicating with a patient to use *swelling* instead of *edema, capable of preventing disease* instead of *prophylactic,* and *heart attack* instead of *myocardial infarction.*

Jargon also refers to abstract, unintelligible, self-important phrasing that conveys an attempt to set oneself apart with special knowledge of a field. Avoid it.

JARGON

➤ Group modalities are effectively utilized in the educational setting during the formative years to restructure behavioral play relationships.

REVISED

➤ Children are taught in school to get along with each other when they play in groups.

To test for jargon, ask yourself, "Have I used the simplest terms available without any loss in communication?"

## 14d Avoiding bias in writing

A bias, or slanted outlook, may be favorable or unfavorable and without good reason either way. Language carries certain biases within it because of the historical circumstances surrounding its development and the ways people have used it. Dominant groups within cultures have often used language to maintain their superior positions and to disempower others. For example, African-Americans have been referred to by such abusive terms as *boy* (adult male), *darky, nigger, spade,* and *blood.* Calling a woman *gal, little woman, the Mrs.,* or *honey* is patronizing and, in certain circumstances, abusive.

Not only can language deny equality; it can deny individuality as well. To assume that all young people are immature or unreasonable or reckless is no better than to assume that all middle-aged people are out of touch or that all people in their later years are senile, infirm, or crotchety. To assume so is to fall prey to stereotypes, to take from people their individuality and uniqueness, and to deny reality.

How can we avoid bias in language? The best way is to cultivate a sense of equality of all people regardless of their age, gender, race, religion, country of origin, or sexual orientation. We must also make an effort not to reinforce stereotypes, intentionally or subconsciously. Edit your writing to eliminate all inadvertent expressions and terms that promote inequality or stereotypes or that may offend your readers by making inappropriate assumptions. More specifically, we can learn how people want to be referred to and respect their wishes. Realize, though, that even the way people wish to be referred to may change—and sometimes rather quickly.

The following list identifies some common sexist terms and alternative usages:

| SEXIST | PREFERRED |
|---|---|
| bellboy | bellhop |
| chairman | chair, head, moderator, chairperson |
| clergyman | member of the clergy |
| congressman | congressional representative |
| fireman | firefighter |
| forefather | ancestor, forebear |
| freshman | first-year student, student |
| landlord/landlady | proprietor, property owner, owner |
| mailman | letter carrier, postal worker |
| male nurse | nurse |
| man, mankind | humans, humankind, humanity, people |
| newsman | reporter, journalist |
| policeman | police officer, officer |
| salesman | sales associate, salesperson |
| stewardess | flight attendant, cabin attendant |
| waitress | waiter, server |

For advice on avoiding sexist pronouns, see page 69.

# *Special Types of Writing*

# 15 Argumentative Essays

A written argument tries to convince a reader to agree with a particular point of view or to take a particular course of action. The writer takes a stand on an issue; provides supporting evidence; appeals logically, emotionally, and ethically to the reader; and carefully uses language to move the reader to agree.

In many ways, writing an argument is like writing any other kind of composition. At the beginning you need to choose a topic and generate ideas. You may sketch an outline and organize your material. After you write a draft, you need to revise and, finally, edit. (See pages 20–22.) At the same time, an argument establishes a clear and forceful thesis statement, provides special kinds of evidence, pays particular attention to logical reasoning, anticipates and attempts to rebut the opposition's ideas, and seeks to establish common ground with the audience. Finally, an argument tries to convince the reader of the validity of the writer's point of view and to establish the writer's *credibility,* the belief that the writer is worth hearing.

## 15a Understanding the elements of argument

There are four main parts to an argument: a topic; claims made about it; data or evidence in support of the claims; and warrants, or assumptions, that link the evidence to the claims. The terms *claim, data,* and *warrants* are based on the work of the English philosopher Stephen Toulmin.

### Topics

When you write an argument, you often know your topic and already have some opinion about it. If, however, you do not already have a subject and a topic, you should review the techniques described earlier for selecting them. (See pages 3–9.) Of course, your opinion should be about a topic people care about, on which there is reasonable disagreement, and which fits the requirements of your assignment.

### Claims

The *claim* of an argument is the opinion you have about your topic. In making a formal claim, you need to take your opinion and make it into a thesis statement (see pages 12–13) that can be supported with data, or evidence:

CLAIM Harry Truman was a great president.

[This statement can never be proved true or false; it remains a judgment. If you provide enough good data, however, you may convince your reader that it is cogent—that it logically follows from the evidence.]

Remember that a claim is an opinion. A fact cannot be a claim because a fact cannot be argued. It is either true or not true.

FACT Harry Truman was the thirty-third president.

[You can check an encyclopedia to find out whether or not Truman was the thirty-third president.]

A belief cannot be a claim because a belief is not based on facts and cannot be tested.

BELIEF The most important quality of a president is common sense.

[A belief looks like a claim but is not based on facts and is therefore not arguable. Some people may believe that Truman's most important quality was not his common sense but his intelligence or his honesty or his directness.]

### Data, or evidence

*Data*, or "reasons" as Toulmin calls them, are the evidence used to support a claim. There are various kinds of data:

*Facts* are statements that can be verified. (*Surinam is located on the northeast coast of South America.*)
*Statistics* are facts expressed in numerical form. (*The violent crime rate has fallen 30 percent over the last five years.*)
*Examples* are specific cases in point. (*Some students, Jaime and Kris for example, could not attend the meeting.*)

*Expert testimony* is the judgment of an authority. (*The U.S. Surgeon General has declared that smoking causes cancer.*)

*Research findings* are results of scientific study. (*Child psychologist Karen Wynn has shown that five-month-old babies know simple arithmetic.*)

The amount and quality of the evidence you use to support your assertions will determine your success in convincing your readers. Your evidence should be:

- accurate (reliable, correctly interpreted and honestly communicated, current),
- complete (full and presented in the proper context),
- meaningful (specific and pertinent),
- representative (reflecting the larger body of information it is drawn from),
- appropriate (neither too sophisticated nor too simple for your subject and audience).

Warrants, or assumptions

A *warrant*, or assumption, links the claim and the data used to support it. Often, the connection between the two is obvious, but at times there is an implied relationship, as in the following example:

| | |
|---|---|
| CLAIM | Safe drivers should avoid drinking alcohol. |
| DATA | Studies that show alcohol impairs judgment. |
| WARRANT | Safe drivers require sound judgment. |

The warrant, or assumption, is that drivers need sound judgment in order to drive safely, so data showing that alcohol impairs judgment are very relevant to the claim. If your audience does not share the warrant that safe driving requires sound judgment, then the data will not have much influence on them. It is important, therefore, that as you write your argument you attempt to determine which warrants you and your audience are likely to accept and which you are likely to reject.

Understanding these terms will help you formulate a sound argumentative thesis. It will also help you see how elements relate, how your argument can be organized, and how faults in reasoning can be detected and corrected.

## 15b Making appropriate appeals

An effective argument can appeal to the reader in three basic ways—through logic, through emotion, and through ethics. Being reasonable is the most important requirement of argumentative writing. Perhaps nothing is more convincing to a reader than a good, clear, logical approach to an issue. However, as humans we are not merely logical machines; our emotional responses to issues are important to us and sway our thinking on issues. When an editorial writer argues for stricter community leash laws by telling about a child bitten by a dog, she appeals to our fears for our own safety. Such fears can change minds. Emotional appeals are not a substitute for sound reasoning but an effective addition.

Finally, ethical appeals are based on the writer's credibility. You establish credibility by being knowledgeable about your subject; by taking the opposition into account; and by maintaining a fair and even tone. Back up your assertions with solid evidence instead of spouting generalities. Use language sincerely and carefully: be aware of connotative meanings, and avoid pretentious words and doublespeak. (See pages 37– 39.) At all times, show respect for the intelligence and interest of your reader.

## 15c Considering your audience

Your audience is those who will read what you write. The more you know about your audience, especially when writing an argumentative essay, the better you will be able to anticipate your readers' reactions to the overall point you are trying to make and to the specific kinds of evidence you are presenting. The Checklist for Questions About Audience is designed to remind you of questions you should ask while writing an argumentative paper.

### Checklist for Questions About Audience

1. Who are my readers?
2. Is my audience specialized (my chemistry lab partners, other Macintosh computer users) or general (literate adults)?

(*continued*)

**Checklist** *(continued)*

3. What do I know about my audience (age; sex; amount of education; religious, social, economic, and political attitudes)?
4. What do my readers know about my subject? What is their knowledge level—expert or novice?
5. What preconceptions do my readers bring to a consideration of the subject? Will they have misconceptions that I can clarify?
6. What do my readers need to know that I can tell them?
7. What is my relationship to my audience: boss? equal? subordinate?
8. How will my audience respond to what I have to say? Will my readers be interested, open-minded, resistant, or hostile?
9. Is there any specialized language that my audience needs or that I should avoid?
10. What do I want my readers to do? How can I help them?

The best writers empathize with their readers. They try to see issues as their readers might; to recognize, understand, and address readers' problems; and to appeal to readers' emotions, rational faculties, and humanity.

## 16 Essay Examinations

In many ways, writing essays in answer to examination questions is similar to other writing tasks—you gather information and formulate ideas, organize them, and express them clearly and coherently. But there is one essential difference with essay examinations: a time constraint.

It is essential that you budget your time. Read and understand the examination questions, and then determine how much time you will spend on each. For example, if you are required to write four essays in sixty minutes and they are weighted equally, you need to spend fifteen minutes per question. If, however, you are asked to answer two short questions, each accounting for fifteen percent of the total grade, you should not spend undue time on the short questions even if you can answer them in great detail. Be sure to allow a little time at the end to read over what you have written and make corrections before turning in your examination.

Your essay must answer the question the instructor has asked. Here is a typical essay examination question for an introductory psychology course:

➤ Define the following concepts as Freud uses them: id, ego, and superego. Explain how these concepts interact.

Notice that the three terms are referred to as *concepts,* a word that may or may not bear special significance in your reading and course. More important, notice that the instructor is asking you to perform two tasks: to define *and* to explain. Planning your essay need not—and should not—take long. Making a short list of the major points you want to be sure to include will keep you from straying off the topic as you write and will help you organize a coherent response.

As in all your writing, your examination essays should demonstrate sound reasoning and an understanding of causal relationships, anticipate opposing arguments and respond to them, and provide evidence and specific detail to support your arguments.

## 17 Response/Reaction Writing

Reading and writing, like listening and speaking, are an interdependent pair of activities. If one-half of the reading/writing pair is not effectively used, the other half of the pair becomes ineffective as well. Therefore, many instructors assign *response/reaction papers* (sometimes referred to as reader/response papers) that ask you to actively engage the text you are reading by examining it critically and writing about your findings. There are many worthwhile purposes for such an assignment:

- to think about what you have read,
- to determine purpose and thesis,
- to evaluate information and evidence to apprehend structure,

- to assess implications,
- to reflect on assumptions the writer makes, and
- to judge overall usefulness.

### 17a How to write a response/reaction essay

Most response/reaction papers are about one page in length. Some instructors call for strictly factual, objective response/reactions, while others allow for various degrees of subjectivity. Your instructor will explain to what degree, if any, you can incorporate a subjective response. Take notes on important features of the text you are reading, and draft and revise as you would any piece of writing. Check on the accuracy of what you are reading by constantly referring back to the text—thus employing both parts of the reading/writing interchange. Be sure to state the writer's thesis, summarize the major points of the work, and relate the work to the body of reading you have already done on the subject. If you are allowed personal comments, assess how convincing the writer's argument is and how valuable you find the author's treatment of the subject.

## 18 Review of Literature Writing

A *review of literature* is common in many disciplines and refers to a special kind of writing in which you classify and evaluate what scholars and researchers have written about a particular topic. Your instructor will give any special instructions, especially on the form your citations should take, for accomplishing your review of literature assignment. In general, two types of information are conveyed in a review of literature: what's available and how good it is. Your task is not simply to list everything you find on a topic and describe what each author has to say. In a review of literature, you must determine which articles and books are most relevant to your topic, analyze them, and then synthesize your findings into a comprehensive and integrated review.

### Checklist for a Review of Literature Essay

- Have I appropriately focused the topic of my review of literature? What question(s) am I trying to answer (e.g., What is the English-only movement all about? What is the focus of recent research on diabetes?)?
- Have I determined the scope of my research? Is my review of literature to be drawn from articles, books, government documents, the popular or online press, or some combination of these sources?
- Have I used the best research tools for this project?
- Have I used both print and electronic databases and materials?
- Have I read all my materials carefully to ensure that each item contains information related to my thesis or research question?
- Have I included materials that take all sides of my research question into account?
- Have I used the appropriate documentation style in my list of citations?

### Checklist for an Item in a Review of Literature

- What is the author's purpose? To present new findings? To argue against accepted beliefs? To present new methods of research and strategies?
- Is the author's thesis or research question clearly stated?
- Can you follow the logic of the author's argument? Does it make sense to you?
- What is the author's theoretical framework? Is the author approaching the research question from a recognizable perspective?
- What is the author's research methodology? Do you think it an appropriate methodology, given the research question?
- Would you draw the same conclusions as the author, given the research findings?

(*continued*)

Checklist *(continued)*

- Is the author's approach to the research problem traditional or innovative? If innovative, is the departure from tradition justified by the findings?
- Does the author acknowledge similar studies carried out earlier?
- Does the author recognize opposing theories, research methods, and results?
- Is the author's research significant? How helpful is it to your understanding of the research problem?

## 19 Annotated Bibliography vs. Abstract

An *annotated bibliography* is a compilation of citations to books, articles, and other documents accompanied by a brief statement or annotation (usually not more than 150 words) that describes and critically evaluates the work. An *abstract,* unlike an annotation, presents an objective description of the work and refrains from any critical judgment. Abstracts are most often found at the beginning of journal articles in the sciences and social sciences, but many instructors assign them as study and review aids.

## 20 Reports

From department to department and instructor to instructor, reports vary greatly in purpose, content, and format, so it is best to consult with your instructor regarding any special requirements that may exist. Whether it's a document reporting on the most effective medicines to treat HIV, the public's attitude toward air travel security, or the latest data on the country's economic strength, most reports contain the following:

- A cover page giving the title of the report, your name, the date, and the person or group for whom the report is written.

- An abstract (see page 50).
- An introduction that makes clear the scope of the project, what you are trying to accomplish with the report, and a discussion of any special terms or circumstances that the reader needs to have for a clear understanding of the report.
- A body that presents your ideas and the information you have gathered. If the report is argumentative, present your thesis or assertion and the evidence necessary to support it. That evidence might be anecdotal or scholarly, oral or written, but it should try to convince your reader.
- A conclusion that summarizes the ideas and information you have presented or that quickly reiterates your argument. The conclusion should, of course, reflect the purpose you presented in your introduction as well as answer any questions you raised.

## 21 E-mail

E-mail offers a fast, convenient, and versatile way to communicate. There are a few general rules for e-mail. Check your inbox regularly, and respond to your messages. When you send an e-mail, use the subject line to give your recipient an immediate indication of what the e-mail contains. If you choose to forward a message from one correspondent to another, make sure the original author does not mind. Do not use all capital letters in any message—it communicates a strident tone and is the equivalent of shouting. Finally, never send an e-mail before double-checking exactly where it is being sent. An e-mail intended to amuse a friend might prove problematic if accidentally sent to an entire listserv or company address list.

The writing and content of personal e-mails is frequently quite informal, which does not present a problem when you send it to your intended recipient only. However, take care not to let such an unconventional style carry over to your more official or widely read e-mail messages.

# 22 Oral Presentations

With time and practice, most people can develop into effective speakers. Good preparation will help to calm any nervousness you may have so you can capture the interest of your audience.

Establish goals for the presentation, taking into account both your subject matter and your audience. Are you arguing a point, presenting information, entertaining your audience, or a combination of all three? What information, emotion, or opinion do you want members of your audience to take away with them?

The best way to prepare for a presentation is to create an outline, prepare and practice, and choose the right visuals.

## 22a Creating an outline

Most presentations have time schedules, which will dictate how in-depth your presentation can be. Create an overall outline that maps out how much time you plan to devote to each component of your presentation. For example, for an hour presentation, you could construct the following outline:

I. Introduction, background information, presentation preview, 10 minutes
II. Discussion of current research by others in the field, 10 minutes
  A. Graphs A and B
  B. Photo slides 1–3
III. Discussion of proposed research by my team, 25 minutes
  A. Graphs C–G
  B. Photo slides 4–10
IV. Conclusion and Q&A, 15 minutes

You may find as you are making your presentation that you wish to change the emphasis somewhat—or audience participation may change it for you—but the outline gives you a solid foundation from which to prepare.

## 22b Preparing and practicing

If you are familiar with your subject matter and have some experience as a speaker, it is usually best to prepare your presentation in note form. Take the strongest points of your presentation, and write them down. For each point, write short reminders about related information that will help you expand on the basic material. Write down facts or figures you need that you may not remember in front of the audience. If you are using visuals for your presentation, such as overheads or slides, you can put the notes right on some of the visuals.

If you think you need more structure than notes give you, you may want to write out the whole presentation. If you do, keep in mind the following points:

- Use short sentences and natural language.
- Format the text for maximum ease of reading. Use large type and at least double-spacing.
- Pay close attention to your transitions. Make sure your audience can readily follow your ideas.
- Do not be afraid to use repetition. Something that is redundant in a written essay will often help audiences follow an oral presentation.
- Craft an effective conclusion.

Practice your presentation. If you have a tolerant friend, practice at least some of the talk in front of him or her. If not, a watch and an empty room will do. Time yourself to make sure you are close to your allotted time—ideally, you should be slightly under to allow for interruptions and questions during the presentation.

If you are using notes, make sure you can flow from point to point without getting confused or searching for information. If you have a prepared text, listen carefully to how it reads. Also, think about how to enliven your delivery and make it more interesting. Monotone droning can undermine even a compelling and well-prepared presentation.

## 22c Choosing the right visuals

All presentations benefit from effective visuals (see pages 138–141 and 171–172). A simple text outline displayed on a screen during your introduction will show your audience the key points of your presentation and foreshadow some of the information you will cover. Charts, graphs, and illustrations not only efficiently convey large amounts of information, they also help structure the presentation and engage the audience's attention.

# *Editing for Grammar*

# 23 Subject–Verb Agreement

In grammar, *subject–verb agreement* refers to the correspondence in form between a verb and its subject. Every verb in a clause or sentence must agree in *number* and *person* with its subject.

SINGULAR THIRD-PERSON SUBJECT AND SINGULAR VERB

➤ *Robert makes* a wonderful pound cake.

PLURAL THIRD-PERSON SUBJECT AND PLURAL VERB

➤ *Hurricanes destroy* millions of dollars worth of property each year.

Normally, we do not think twice about subject–verb agreement. A few constructions, however, can be troublesome.

## 23a Words between subject and verb

Sometimes intervening word groups obscure the relationship between the subject and the verb in a sentence. A prepositional phrase is usually the culprit.

➤ Three movies in that cineplex ~~deals~~ deal with natural disasters. [The plural subject *movies* takes a plural verb.]

Another word group that can cause confusion about subject–verb agreement is the subordinate clause.

➤ Taking mood-altering drugs, although the drugs sometimes provide great medical benefits, ~~are~~ is generally criticized by the public. [The singular subject *taking* takes a singular verb.]

Expressions such as *along with, as well as, in addition to, including,* and *together with* do not make a singular subject plural.

➤ The cost of medication as well as doctors' fees ~~are~~ is a problem for retired people. [The singular subject *cost* takes a singular verb.]

### 23b Subjects joined by *and*

In most cases, compound subjects take a plural verb.

➤ Their television and their camera ~~was~~ were stolen.

However, when singular subjects joined by *and* are preceded by *each, every, no, nothing,* or *many a(n),* use a singular verb.

➤ Every chair and every couch ~~are~~ is on sale this week.

If the two parts of a compound subject, whether singular or plural, express a single idea or refer to a single person, use a singular verb.

➤ Ham and eggs ~~are~~ is my favorite breakfast.

### 23c Subjects joined by *or* or *nor*

With compound subjects, make the verb agree with the subject that is closest to it.

➤ Neither the instructor nor the students ~~is~~ are satisfied with the classroom. [One singular subject (*instructor*) and one plural subject (*students*) are joined by *nor;* make the verb agree with the closer subject (*students,* which is plural).]

➤ Either Mary or I ~~has~~ have to report the broken window. [One third-person subject (*Mary*) and one first-person subject (*I*) are joined by *or;* make the verb agree with the closer subject (*I,* which is first person).]

### 23d *Family* and other collective nouns

Collective nouns are singular in form but can have a plural meaning—for example, *band, jury, family, minority, majority.* They may take either a singular or a plural verb, depending on whether you are considering the group as a single unit or in terms of its multiple members. To be consistent in sentences containing collective nouns, make pronoun references to the noun singular or plural as well.

➤ The *committee gives its* report today. [*Committee* is considered as a singular subject here because all the members are acting as one in giving a single report; a singular verb is used together with a singular pronoun.]

➤ The *committee give their* reports today. [The multiple members of the committee will each give a report; *committee* here has a plural meaning, so the verb and pronoun reference must be plural.]

## 23e *Who, which,* and *that*

Usually, verb agreement is not a problem with relative pronouns.

➤ A *politician who votes* for labor gets elected. [The singular noun *politician* is the antecedent of *who;* use a singular verb to agree with that antecedent.]

➤ *Politicians who vote* for labor get elected. [The antecedent of *who* is *politicians,* which is plural; use a plural verb to agree with that antecedent.]

However, when the phrase *one of the* or *the only one of the* is part of the sentence, identifying the antecedent takes extra care.

➤ She is one of the *senators who* always vote for labor. [The antecedent of *who* is *senators* because the sense is that more than one senator votes a particular way; use a plural verb to agree with that antecedent.]

➤ She is the only *one* of the senators *who* always votes for labor. [The antecedent of *who* is *one* because the sense is that *this particular senator* stands out from the others; use a singular verb to agree with that antecedent.]

## 23f *Anybody* and other indefinite pronouns

Indefinite pronouns such as *anybody, anyone, anything, each, either, everybody, everyone, everything, neither, no one, one, somebody, someone,* and *something* are singular and take singular verbs.

➤ Each of the students ~~have~~ has to pick a major.

With indefinite pronouns that can be either singular or plural (*all, any, none, some*), use a singular or plural verb, depending on the word that the indefinite pronoun refers to.

➤ *All* of the *wheat is* harvested.

➤ *All* of the *fields are* plowed. [Because *wheat* is singular, *all* takes a singular verb. Because *fields* is plural, *all* takes a plural verb.]

## 23g Subject after verb

When you invert normal word order, be especially careful to identify the subject correctly.

➤ From his research ~~has~~ have come three promising cancer *drugs*. [The plural subject *drugs* takes a plural verb.]

With sentences beginning with *there* or *here*, the subject follows the verb.

➤ There ~~is~~ are stone, brick, and tile fireplaces in Blenheim Castle.

## 23h Linking verbs with subjects and subject complements

A subject complement renames or describes the subject but is not the subject. Make sure your verb agrees with the subject, not with the subject complement.

➤ His main interest ~~are~~ is stamps. [The subject, *interest*, is singular, so the verb must be singular; *stamps* is the subject complement.]

## 23i *Statistics* and other singular nouns ending in *-s*

Nouns that are plural in form but singular in meaning take singular verbs. Examples include *athletics, economics, mathematics, mumps, news, physics,* and *politics.*

➤ Measles ~~were~~ was a serious childhood disease until a vaccine was developed.

Some nouns, however, end in *-s* and are singular in meaning but take plural verbs: *eyeglasses, pants, pliers, scissors, trousers, tweezers.*

➤ The scissors ~~is~~ are on the table.

If you are uncertain whether to use a singular or a plural verb with a particular noun, consult your dictionary.

### 23j Titles and words used as words

➤ *The Canterbury Tales* ~~are~~ is by Chaucer.

➤ *Octopi* ~~are~~ is one of the plural forms of *octopus.*

## 24 Verbs: Form, Tense, Mood, and Voice

### 24a Irregular verbs

English verbs are either *regular* or *irregular* in form. A regular verb forms both the past tense and the past participle by adding *-ed* to the base, or infinitive, form (*walk, walked, walked*). Regular verbs rarely cause problems; it is irregular verbs that can be tricky because they have different forms for the past tense and the past participle, as the following sentences illustrate.

➤ Americans usually *sing* the national anthem before sporting events.

➤ Mariah Carey *sang* the national anthem at the football game.

➤ Garth Brooks had *sung* the anthem the year before.

The most frequently used of the approximately two hundred irregular verbs in English are identified in the following list. For some, two acceptable forms are given; they usually represent regional variations.

| INFINITIVE | PAST TENSE | PAST PARTICIPLE |
|---|---|---|
| awake | awoke | awakened |
| be | was | been |

| INFINITIVE | PAST TENSE | PAST PARTICIPLE |
|---|---|---|
| bear | bore | borne |
| become | became | become |
| begin | began | begun |
| bend | bent | bent |
| bet | bet | bet |
| bite | bit | bitten, bit |
| blow | blew | blown |
| break | broke | broken |
| bring | brought | brought |
| build | built | built |
| burst | burst | burst |
| catch | caught | caught |
| choose | chose | chosen |
| come | came | come |
| cost | cost | cost |
| cut | cut | cut |
| deal | dealt | dealt |
| dig | dug | dug |
| dive | dived, dove | dived |
| do | did | done |
| drag | dragged | dragged |
| draw | drew | drawn |
| dream | dreamed, dreamt | dreamed, dreamt |
| drink | drank | drunk |
| drive | drove | driven |
| eat | ate | eaten |
| fall | fell | fallen |
| feel | felt | felt |
| find | found | found |
| fit | fit, fitted | fit, fitted |
| fly | flew | flown |
| forbid | forbade, forbad | forbidden |
| forget | forgot | forgotten, forgot |
| freeze | froze | frozen |
| get | got | got, gotten |
| give | gave | given |
| go | went | gone |
| grow | grew | grown |
| hang (to suspend) | hung | hung |
| hang (to execute) | hanged | hanged |
| hear | heard | heard |
| hit | hit | hit |
| hurt | hurt | hurt |

| INFINITIVE | PAST TENSE | PAST PARTICIPLE |
|---|---|---|
| know | knew | known |
| lay (to put) | laid | laid |
| lead | led | led |
| lend | lent | lent |
| let | let | let |
| lie (to recline) | lay | lain |
| light | lighted, lit | lighted, lit |
| lose | lost | lost |
| pay | paid | paid |
| put | put | put |
| ride | rode | ridden |
| ring | rang | rung |
| rise | rose | risen |
| run | run | run |
| say | said | said |
| see | saw | seen |
| set (to place) | set | set |
| shake | shook | shaken |
| shine | shone, shined | shone, shined |
| shrink | shrank | shrunk |
| shut | shut | shut |
| sink | sank | sunk |
| sit (to be seated) | sat | sat |
| slay | slew | slain |
| speak | spoke | spoken |
| spread | spread | spread |
| spring | sprang | sprung |
| steal | stole | stolen |
| strike | struck | struck, stricken |
| swear | swore | sworn |
| swim | swam | swum |
| take | took | taken |
| teach | taught | taught |
| tear | tore | torn |
| throw | threw | thrown |
| wake | woke, waked | waked, woken |
| wear | wore | worn |
| win | won | won |
| write | wrote | written |

If you are not certain about the form of a particular verb, consult your dictionary. Most dictionaries list all three principal parts of irregular verbs and the infinitive form of regular verbs.

## 24b *Lay* and *lie; set* and *sit*

Writers often confuse the irregular verbs *lay* (to put down) and *lie* (to recline) and *set* (to put something somewhere) and *sit* (to take a seat).

| INFINITIVE | PAST TENSE | PAST PARTICIPLE |
|---|---|---|
| lay | laid | laid |
| lie | lay | lain |
| set | set | set |
| sit | sat | sat |

*Lay* and *set* are transitive verbs and require a direct object (DO) to complete their meaning. *Lie* and *sit* are intransitive verbs and therefore do not take a direct object.

➤ I laid the package (DO) down on the table in the hall yesterday.

➤ I'm going to lie down for an hour before dinner.

➤ You can set the groceries (DO) on the kitchen counter.

➤ Ms. Lane asked the class to sit quietly.

## 24c Verb tenses

The tense of a verb indicates when an action is taking place. In English, there are three basic tenses: present, past, and future. Each tense also has a perfect, a progressive, and a perfect progressive form. Use the present tense to express an action occurring in the present and for actions that occur regularly. Use the past tense to describe an action that occurred entirely in the past. Use the future tense to describe an action that will take place in the future. The basic tenses for the regular verb *talk,* the irregular verb *break,* and the very irregular verb *be* follow.

**PRESENT TENSE**

***singular***

| | |
|---|---|
| I | talk, break, am |
| you | talk, break, are |
| he/she/it | talks, breaks, is |

*plural*

| | |
|---|---|
| we | talk, break, are |
| you | talk, break, are |
| they | talk, break, are |

PAST TENSE

*singular*

| | |
|---|---|
| I | talked, broke, was |
| you | talked, broke, were |
| he/she/it | talked, broke, was |

*plural*

| | |
|---|---|
| we | talked, broke, were |
| you | talked, broke, were |
| they | talked, broke, were |

FUTURE TENSE

| | |
|---|---|
| I, you, he/she/it, we, they | *will* talk, break, be |

The perfect tenses are used to indicate more complex actions. The present perfect tense describes an action that occurred in the past but continues into or affects the present. The past perfect describes an action that preceded another action when both occurred in the past. The future perfect describes an action that will be completed before another future event. You can create all three perfect tenses by using a form of *have* plus the past participle.

PRESENT PERFECT

| | |
|---|---|
| I, you, we, they | *have* talked, broken, been |
| he/she/it | *has* talked, broken, been |

PAST PERFECT

| | |
|---|---|
| I, you, he/she/it, we, they | *had* talked, broken, been |

FUTURE PERFECT

| | |
|---|---|
| I, you, he/she/it, we, they | *will have* talked, broken, been |

The progressive forms of each of the six tenses discussed here are used to express ongoing action. You can create a progressive verb by using a form of *be* together with the present participle.

PRESENT PROGRESSIVE

| | |
|---|---|
| I | *am* talking, breaking, being |
| he/she/it | *is* talking, breaking, being |
| you, we, they | *are* talking, breaking, being |

PAST PROGRESSIVE

| | |
|---|---|
| I, he/she/it | *was* talking, breaking, being |
| you, we, they | *were* talking, breaking, being |

FUTURE PROGRESSIVE

| | |
|---|---|
| I, you, he/she/it, we, they | *will be* talking, breaking, being |

PRESENT PERFECT PROGRESSIVE

| | |
|---|---|
| I, you, we, they | *have been* talking, breaking, being |
| he/she/it | *has been* talking, breaking, being |

PAST PERFECT PROGRESSIVE

| | |
|---|---|
| I, you, he/she/it, we, they | *had been* talking, breaking, being |

FUTURE PERFECT PROGRESSIVE

| | |
|---|---|
| I, you, he/she/it, we, they | *will have been* talking, breaking, being |

**The literary or historical present tense.** Use the present tense for events that have occurred in the past for the purpose of making the events or fictional happenings come alive. Do not lapse into the past tense out of habit because you are writing about something that has already happened in your reading.

➤ Tragically, King Lear ~~trusted~~ trusts Goneril and Regan.

**Present tense for future action.** You can use the present tense to indicate actions that will take place at some future time.

➤ Fran *leaves* for Chapel Hill on Saturday.

## 24d Consistency and sequences of tenses

You must show a logical relationship between the time expressed by verbs in a main clause and the time expressed by verbs in subordinate clauses, infinitives, or participles.

➤ The tiger jumped through the hoop when the trainer ~~gives~~ gave the command.

Mixing tenses does not make a logical statement: you cannot have the tiger jumping in the past tense and the trainer giving the command in the present. Put both verbs in the past tense to show that both actions already happened, or put both verbs in the present tense to indicate a habitual action.

### Subordinate clauses

You may use any verb tenses in your main clause (MC) and in your subordinate clause (SC) that are compatible with the intended meaning of your sentence.

➤ [Some optimistic economists *believe*]MC [that the stock market soon *will top* last year's record.]SC

However, when the verb in the main clause is in the past or past perfect tense, you must also use the past or past perfect tense in the subordinate clause.

➤ [The Allies *had put* the battle out of reach]MC [before the Germans *changed* their strategy.]SC

EXCEPTION When a subordinate clause presents a general truth, use the present tense even though the verb in the main clause is in the past or past perfect.

➤ [We *learned* in business ethics]MC [that power *corrupts*.]SC

### Infinitives

Use the present infinitive (*to praise, to drink*) to express action that occurs at the same time as or later than that of the other verb in the sentence.

➤ Marta tried *to improve* her tennis game.

Use the present perfect infinitive (*to have praised, to have drunk*) to express action that occurs earlier than that of the other verb in the sentence.

➤ I would like *to have attended* the Alicia Keyes concert last weekend.

### Participles

Use the present participle (*praising, drinking*) to express action happening at the same time as that of the other verb in the sentence.

➤ *Flying* across the Atlantic, Lindbergh never lost his confidence.

Use the present perfect participle (*having praised, having drunk*) for action occurring earlier than that of the other verb in the sentence.

➤ *Having answered* questions for an hour, the president concluded the news conference.

## 24e Mood

The three moods of a verb—indicative, imperative, and subjunctive—reveal the writer's intent or how the writer views a thought or action. For example, it is a verb's mood that shows the writer is making a statement of fact and not simply expressing a wish.

Use the indicative mood to state a fact or opinion or to ask a question. Use the imperative mood to give an order.

➤ *Gargle* with saltwater.

➤ *Be* at the airport by noon.

Use the subjunctive mood in *that* clauses expressing demands, resolutions, or requests and to express a condition contrary to fact or a wish.

➤ The manager asked that he *list* his qualifications for the job.

➤ If I *were* younger, I would go to law school.

## 24f Voice

Voice indicates whether the subject of a sentence is the actor or the receiver of the action. In the active voice, the subject of the sentence does the acting; in the passive voice, the subject is acted upon.

ACTIVE VOICE

➤ The National Rifle Association *opposes* gun-control legislation.

PASSIVE VOICE

➤ Gun-control legislation is opposed by the National Rifle Association.

Because the active voice is more concise and more direct than the passive voice, writers—and readers—prefer it. To change a sentence from passive to active voice, simply make the actor the subject of the sentence and eliminate unnecessary words.

The students selected
➤ ^"Poverty and Homelessness" ~~was selected~~ as the theme for next year's lecture series^. ~~by the students.~~

When the action itself is what is important or you wish to minimize the importance of the actor, use the passive voice. The passive voice is often used in scientific and technical writing to emphasize processes and events rather than individuals.

➤ Three drops of boric acid *were added* to the solution.

➤ Investigative teams *were sent* by the Centers for Disease Control.

# 25 Pronoun Problems

A pronoun takes the place of a noun in a sentence. Pronoun problems fall into three general categories: problems with agreement (a pronoun must agree with the noun to which it refers), referent problems (a pronoun must clearly refer to another word), and case problems (a writer must know when to use *I* and when to use *me,* when to use *who* and when to use *whom*). The issues of agreement, reference, and case can produce sexist language, lack of clarity, or an impression of carelessness.

## 25a Pronoun–antecedent agreement

A pronoun must agree with its antecedent, the word to which the pronoun refers, in gender, number, and person.

### Sexist pronouns

Traditionally, writers have used a masculine, singular pronoun to agree with singular indefinite antecedents (such as *anyone, someone,* and *everyone*) and to refer to generic antecedents (such as *employee, student, patient, traveler,* or *applicant*). But *anyone* can be female or male, and women are employees (or students, patients, travelers, applicants), too. Using masculine pronouns to refer to both females and males is unacceptable because it is sexist; that is, such usage leaves out women as a segment of society or diminishes their presence. Instead, use *he or she, his or her.* Sometimes the best solution is to rewrite the sentence to put it in the plural or to avoid the personal pronouns altogether.

**UNACCEPTABLE BECAUSE IT IS SEXIST**

➤ If *anyone* wants a season ticket, *he* will have to sign up before Friday.

**UNACCEPTABLE BECAUSE IT IS GRAMMATICALLY INCORRECT**

➤ If *anyone* wants a season ticket, *they* will have to sign up before Friday.

**REWRITTEN USING *HE OR SHE***

➤ If *anyone* wants a season ticket, *he or she* will have to sign up before Friday.

**REWRITTEN IN THE PLURAL**

➤ If *students* want season *tickets, they* will have to sign up before Friday.

**REWRITTEN TO AVOID THE PRONOUNS**

➤ All season tickets must be reserved before Friday.

**REWRITTEN IN THE SECOND PERSON**

➤ If *you* want a season ticket, *you* must sign up before Friday.

**REWRITTEN WITH A RELATIVE PRONOUN**

➤ *Anyone* who wants a season ticket must sign up before Friday.

**NOTE** Disregard gender and use a plural pronoun to refer to two or more singular antecedents that are connected by *and.*

➤ Dinah and Jack were exhausted after ~~her and his~~ their chemistry final.

### Collective nouns as antecedents

If a collective noun (such as *family, majority, choir, jury, faculty*) is understood as a unit, it takes a singular pronoun; if it is understood in terms of its multiple members, it takes a plural pronoun.

➤ The band started to play ~~their~~ its first number.

➤ The band agreed to pay for ~~its~~ their own uniforms.

### Antecedents with *or* or *nor*

When two or more antecedents are connected by *or* or *nor,* use a pronoun that agrees with the closest antecedent.

➤ Neither the judge nor the lawyers were willing to change ~~his~~ their position.

### *Who, whom, that,* and *which*

Use *who* or *whom* to refer to persons; use *that* and *which* to refer to animals and things.

➤ The bicyclist swerved to avoid the dog ~~who~~ that had wandered into the street.

It is acceptable to use *who* with an animal that has a name, however.

➤ Burt is a friendly dog ~~that~~ who likes to bark.

## 25b Pronoun reference

To avoid repeating nouns in our speech and writing, we use pronouns as noun substitutes. The noun for which a pronoun stands is called its antecedent, or referent.

➤ As Laurie opened the envelope, ~~Laurie~~ she held ~~Laurie's~~ her breath.

When the relationship between a pronoun and its antecedent is unclear, the message is inaccurate or ambiguous (has two or more meanings). The more words that intervene between the antecedent and the pronoun, the more chance there is for confusion.

### Ambiguous reference

If a pronoun can refer to either one of two antecedents, ambiguity exists. Either repeat the antecedent, or rewrite the sentence.

➤ Liz told her mother that her mother's sweater had a hole in it.

➤ Liz said to her mother, "Your sweater has a hole in it."

Be especially careful that ambiguity does not result when using *it* as a pronoun referent.

➤ When Alex drove the car through the garage door, he badly damaged the car ~~it~~.

### Implied reference

Every time you use a pronoun in a sentence, you should be able to point to its noun equivalent. If you cannot, use a noun instead of the pronoun.

➤ Because the hurricane was a potential danger to coastal communities, inhabitants ~~they~~ were told to evacuate.

Sometimes a word or phrase, modifier, or possessive that implies a noun is mistaken for an antecedent.

➤ In ~~Langston Hughes's~~ "Salvation," Langston Hughes ~~he~~ tells how his family tried to save him from sin when he was a boy.

### Vague reference

Use *this, that, which,* and *such* judiciously to refer to a general idea in a preceding clause or sentence. If an idea is relatively simple, no confusion results from such a construction.

➤ I lost five pounds, and this made me happy.

If the pronoun refers to a broader or more general idea, however, vagueness can result. To correct the problem, either substitute a noun for the pronoun, or provide an antecedent to which the pronoun can clearly refer.

➤ One plumber supervised while the other one did the work, an arrangement that ~~which~~ is normal.

The pronouns *it, they,* and *you* must have an antecedent. In speech, we use *it, they,* and *you* loosely, without specific antecedents, and our listeners do not mind. But in written English, when these pronouns lack an antecedent, the writer gives an impression of vagueness and carelessness.

➤ ~~On the news, it~~ The newscaster said the earthquake hit 6.5 on the Richter scale.

➤ In the book, ~~they wrote that~~ Coach is a cocaine addict.

➤ If the driver is not careful, ~~you~~ he or she can easily miss the exit for San Jose.

## 25c Case of personal pronouns

In English, the case, or form, of a personal pronoun depends on the grammatical function the pronoun serves in a sentence.

| SUBJECTIVE CASE | OBJECTIVE CASE | POSSESSIVE CASE |
|---|---|---|
| I | me | my, mine |
| you | you | your, yours |
| he/she/it | him/her/it | his/her/its, his/hers/its |
| we | us | our, ours |
| they | them | their, theirs |
| who | whom | whose |

### Pronouns as subjects and subject complements

Use the subjective case of pronouns for subjects and subject complements. Subject complements usually follow a form of the verb *be.*

➤ Marge and ~~me~~ I shared expenses on the trip to Mexico.

➤ The young man Maria dated is ~~him~~ he.

If you think being grammatically correct sounds unnatural, rewrite the sentence to put the pronoun in the subject position.

➤ He is the young man Maria dated.

**Pronouns as direct objects, indirect objects, and objects of prepositions**

Use the objective case for pronouns functioning in these ways.

➤ The dean gave Miguel and ~~I~~ me an hour-long interview.

➤ The professor asked Miguel and ~~I~~ me questions that we were prepared to answer.

➤ Between you and ~~I~~ me, Sharon's plan is best.

**Pronouns to show ownership**

When a pronoun precedes a noun, use the first form of the possessive case (*my, your, his, her, its, our, their, whose*) to show ownership.

➤ *My* notebook has a red cover.

Use the second form of the possessive case (*mine, yours, his, hers, ours, theirs*) when the pronoun follows a verb and does not precede a noun.

➤ The notebook with the red cover is *mine.*

**Pronouns with gerunds**

Pronouns that modify a gerund take the first form of the possessive case (*my, your, his, her, its, our, their, whose*).

➤ ~~Him~~ His returning to college showed great maturity.

➤ The car salesperson understood ~~us~~ our wanting a good deal.

**Pronouns in compound constructions**

To choose the correct pronoun in a compound construction, identify its function in the sentence: is it a subject or object? Test your choice by mentally blocking off the entire compound construction *except* the pronoun in question.

➤ My three housemates and ~~me~~ I do the cooking. [Read aloud: *Me do the cooking.* You can hear the error; revise to put the pronoun in the subjective case, *I.*]

➤ The French Society appointed Karen and ~~I~~ me co-chairpersons for the coming term. [Read aloud: *The society appointed I.* Again, you can hear the error; revise to use the objective case.]

**Pronouns in appositives**

Because an appositive has the same grammatical function as the noun or pronoun it explains, describes, or identifies, the appositive must be in the same case.

➤ Two members of the debate team, Sara and ~~me~~ I, received certificates. [*Sara and I* is supposed to be in apposition to the subject, *two members;* use the subjective case.]

➤ Let's you and ~~I~~ me leave work early and go surfing. [*You and me* is supposed to be in apposition to *us,* the object of the verb *let;* use the objective case.]

**Pronouns in *as* and *than* comparisons**

When making comparisons using *as* or *than,* writers sometimes omit words. One way to choose the right pronoun is to supply the missing words to determine whether the pronoun is the subject or the object of the implied clause.

➤ Philip wears sweaters more than ~~me~~ I. [The complete understood clause is *than I wear sweaters.*]

➤ My father gives my younger sister as big an allowance as ~~I~~ me. [The complete understood clause is *as he gives me.*]

**Pronouns with infinitives**

Infinitives (*to* plus a verb) present an exception to the rule that the subjective case is used for subjects: subjects as well as objects of infinitives are always in the objective case.

➤ Mother asked Tim and ~~I~~ me to peel the potatoes. [*Me* is the subject of the infinitive, *to peel.*]

➤ My sister named her new baby Leslie, so I expected ~~he~~ him to be a ~~she~~ her. [*Him* is the subject of the infinitive, *to be,* and *her* is its object; the objective case is correct in each instance.]

### *Who* or *whom*

As interrogative pronouns, *who* and *whom* commonly begin questions. But which one should you use? To determine the case of the interrogative pronoun, determine its function in the clause. Sometimes it helps to recast the question in your head or to answer the question.

➤ ~~Whom~~ who among the generals led the Allied forces on D-Day? [Answer the question: he led the forces, not *him* led the forces, so use the subjective case, *who.*]

➤ ~~Who~~ whom is the detective speaking to now? [Answer the question: the detective is speaking to *him* or *her,* so use the objective case, *whom.*]

As relative pronouns, *who, whoever, whom,* and *whomever* introduce subordinate clauses. The case of the pronoun depends on its function within the subordinate clause, regardless of the clause's function in the sentence as a whole.

➤ He did not remember ~~whom~~ who was president in 1976.

➤ The Barnes Prize is awarded to ~~whomever~~ whoever has the highest grade-point average in physics. [Here, the relative pronoun is the subject of the verb *has,* so the subjective case is required—*whoever.* The entire subordinate clause is the object of the preposition *to.*]

➤ Frank said he would select ~~whoever~~ whomever he wanted.

Do not be confused by expressions such as *I believe, she thinks,* and *we know* that frequently follow *who* or *whom* in a subordinate clause.

➤ I plan to call only the people ~~whom~~ who we know will want to come.

## 26 Adjectives and Adverbs

Adjectives and adverbs are modifiers; that is, they limit or qualify the meaning of other words. There are, however, important differences between the two, and they cannot be used interchangeably.

## 26a Adverbs

To avoid inadvertently using an adjective when an adverb is called for, pay attention to what you are modifying. If you are modifying a verb (V), an adjective (ADJ), or an adverb (ADV), use an adverb.

V ADV

➤ The gold dome *shone brightly* [not *bright*] in the dazzling sunshine.

ADV ADJ

➤ Make sure to see her *really* [not *real*] *unusual* collection of masks.

ADV ADV

➤ Gary runs *incredibly fast* for a big man.

## 26b Adjectives

Because the subject of a sentence is always a noun (N) or noun substitute (e.g., PRON) and only adjectives—not adverbs—modify nouns, use an adjective after a linking verb (LV) to modify, or complement, a subject.

N ADJ

➤ The *photography* in the new Audubon movie is *spectacular.*

Commonly used linking verbs include *be, become, appear, grow, seem, remain,* and *prove.* The verbs *feel, look, smell,* and *taste* can function as either linking verbs or action verbs, and therein lies a potential problem.

➤ The diplomat feels ~~badly~~ bad about the stalled peace talks.

➤ The pear salad tasted ~~well~~ good.

If you are describing a state of being, you must use an adjective. If you are describing an action, use an adverb to modify the verb.

N LV ADJ

➤ The *dentist looked curious.* [In this sentence, the dentist appears inquisitive; *looked* is a linking verb, and the adjective *curious* describes his state of being.]

V ← ADV

➤ The dentist *looked curiously* at my chipped tooth.
[Here, the dentist is performing an action. He is looking at something in a curious way; *looked* is an action verb, and an adverb modifies it.]

## 26c Demonstrative adjectives with *kind of, sort of,* and *type of*

With phrases such as *kind of, sort of,* and *type of,* the demonstrative adjective modifies the noun *kind, sort,* or *type* and so must agree with it in number.

➤ ~~These~~ This sort of question comes up during orientation every year.

➤ These kind^s of issues must be dealt with.

## 26d Comparative and superlative forms of adjectives and adverbs

Adjectives and adverbs have three forms: the positive (used to describe: *big, vigorously*), the comparative (used to compare two things: *bigger, more vigorously*), and the superlative (used to compare three or more things: *biggest, most vigorously*).

With few exceptions, one-syllable and many two-syllable adjectives form the comparative by adding *-er* and the superlative by adding *-est*. Adjectives with three or more syllables, as well as some two-syllable adjectives, use *more* and *most,* or *less* and *least,* to form the comparative and the superlative, respectively.

Most adverbs form the comparative and superlative by adding *more* and *most* (or *less* and *least*); a few short adverbs add *-er* or *-est*.

➤ Which of the two candidates is ~~most~~ more qualified in your estimation?

➤ Of the three athletes, the javelin thrower performed ~~more~~ most carefully.

Be alert not to use *more* or *most* (or *less* or *least*) after already adding *-er* or *-est* to the adjective or adverb in question.

➤ According to college polls, the ~~most~~ sexiest actor is Brad Pitt.

➤ Tuesday was all the ~~more~~ gloomier because I failed a pop quiz.

Several adjectives, because of their meaning, do not make sense as comparatives or superlatives. Among them are *perfect* and *unique;* that is, a thing either is or is not unique or perfect.

➤ That's the most ~~unique~~ unusual ring I ever saw.

➤ Her ring is ~~very~~ unique.

## 27 Sentence Fragments

A *sentence fragment* is a part of a sentence presented as if it were a complete sentence. That is, it begins with a capital letter and ends with a period, question mark, or exclamation point, but it does not include one or both of the two essential elements required of a grammatically complete sentence—a subject and a verb. A sentence fragment gives readers a "fragment" of a thought, not a complete thought. Usually, fragments occur unintentionally and are correctable; under certain circumstances, however, you may want to use a fragment intentionally.

You can eliminate sentence fragments in one of two ways: (1) join the fragment to a sentence nearby, or (2) develop the fragment itself into a complete sentence.

### 27a Phrase fragments

A phrase fragment lacks a subject, a complete verb, or both.

➤ We left Rome and traveled northward to the picturesque little hill towns of Italy. ~~Also~~ We also went to Florence and Venice.

Often a verbal—an infinitive, gerund, or participle—is mistaken as the verb.

➤ Fourteen of us went on the desert survival trip. ~~Hiking~~ , hiking into Death Valley and traveling only after dark.

An appositive phrase identifies or explains another noun or noun equivalent and cannot stand alone as a sentence.

➤ Marco read two good books during his vacation. ^, Scott O'Dell's *Island of the Blue Dolphins* and N. Scott Momaday's *House Made of Dawn.*

## 27b Subordinate clause fragments

A *subordinate clause* contains both a subject and a predicate, but it begins with a subordinator, either a subordinating conjunction (e.g., *after, although, because, rather than, as if*) or a relative pronoun (*that, which, who, whom, what*), and therefore is a fragment. To correct a subordinate clause fragment, join it to a nearby sentence, or rewrite it as a sentence itself.

➤ The president met with his staff every week. ~~So~~ , so that problems were rarely ignored.

➤ The president met with his staff every week. ^; ~~So that~~ problems were rarely ignored.

## 27c Intentional fragments

Sentence fragments are not always wrong. In fact, if not overused, sentence fragments can add emphasis.

SENTENCES

➤ Would I change my major now? No, I wouldn't.

DELIBERATE FRAGMENTS

➤ Change my major now? No!

If you are in doubt about the acceptability of a sentence fragment, remember that you will never be wrong in using a complete sentence.

# 28 Comma Splices and Run-on Sentences

A *comma splice,* or comma fault, occurs when a writer uses only a comma or a comma and a conjunctive adverb to join two or more main clauses.

COMMA SPLICES

➤ The legislators could not agree on the bill, they had to stay in session.

➤ The legislators could not agree on the bill, therefore, they had to stay in session.

A run-on sentence, or fused sentence, occurs when no punctuation and no coordinating conjunction appear between two or more main clauses.

RUN-ON

➤ The legislators could not agree on the bill they had to stay in session.

There are four main ways to correct comma splices and run-ons.

1. Make the main clauses into separate sentences.
2. Use a semicolon.
3. Use a comma and a coordinating conjunction.
4. Subordinate one main clause to the others.

An example of each remedy follows.

1. The legislators could not agree on the bill. They had to stay in session.
2. The legislators could not agree on the bill; they had to stay in session.
3. The legislators could not agree on the bill, so they had to stay in session.
4. Because the legislators could not agree on the bill, they had to stay in session.

## 28a Making separate sentences with a period

When the ideas expressed in the two main clauses are not closely related, make each clause into a separate sentence.

➤ Our backpacking trip will be through sparsely settled country. We'll ~~we'll~~ take pictures of the animals we see.

## 28b Connecting clauses with a semicolon

When the ideas expressed in the two main clauses are close and the relationship between them is obvious without a coordinating conjunction, use a semicolon to join them.

➤ The delegates to the convention were disorganized~~,~~; they needed someone to lead them.

## 28c Connecting clauses with a comma and a coordinating conjunction

If the two ideas are equally important and you want to avoid two short sentences, use a comma and a coordinating conjunction.

➤ Many countries signed the treaty~~,~~, but the United States did not.

## 28d Making a subordinate clause or phrase

This method of correction is often the best choice because it offers the greatest variety of possible revisions. You can choose the subordinating conjunction that conveys exactly what you mean and puts the emphasis where you want it in the sentence.

➤ Because the ~~The~~ fog came on suddenly, we couldn't find the dock.

# 29 Common ESL Problems

Mastering a language, whether it is your original or a second language, is an ongoing process. This section is designed to help students of English as a second language (ESL) with the most frequent problems in writing grammatically correct, idiomatic English.

## 29a Nouns, quantifiers, and articles

Two important categories of nouns are noncount nouns and count nouns. *Noncount nouns* name things that cannot be counted separately, such as *air* and *water*. *Count nouns* name things that can be counted separately, such as *students* and *chairs*. This section also gives help in using quantifiers (such as *a little* and *a few*), the indefinite article *a/an*, and the definite article *the*.

### Identifying noncount nouns

Most noncount nouns fall into one of several types.

**NAMES OF FOODS** beef, bread, cauliflower

**LIQUID, POWDERS, GRAINS** milk, dust, sand

**AREAS OF KNOWLEDGE** biology, economics, history

**IDEAS AND EMOTIONS** beauty, hatred, truth

**OTHER NONCOUNT NOUNS** (These are not obvious, so you must learn them.) clothes, furniture, information, machinery, money, news

Do not make noncount nouns plural.

➤ I checked my ~~luggages~~ luggage on the plane.

A partitive is a phrase that indicates the units of something—for example, *a piece of* luggage, *a pound of* sugar, *a gallon of* gasoline. When you use a partitive with a noncount noun, keep the noun itself singular, even when the partitive is plural.

➤ I had two pieces of luggage~~s~~ with me.

Do not use the indefinite article *a/an* with noncount nouns. Use no article when the noun has a general meaning.

➤ ~~A gasoline~~ Gasoline becomes more expensive during holiday weekends.

Use *some* or *the* when the noun has a specific identity that is known to your reader.

➤ I bought ~~a~~ some gasoline yesterday.

➤ The milk is in ~~a~~ the fridge.

### Quantifiers for noncount and count nouns

A *quantifier* is an adjective, a word or phrase that tells how much or how many of something. It takes the place of an article. Some quantifiers work with both noncount and count nouns.

| NONCOUNT NOUNS | COUNT NOUNS |
|---|---|
| *a little* salt | *a few* students |
| *too much* poverty | *too many* cars |
| *a great deal of* confidence | *three* instructors |
| *enough* money | *enough* quarters |
| *some* furniture | *some* chairs |
| *a lot of* bread | *a lot of* oranges |

To ask questions regarding quantity, use "How much?" with noncount nouns and "How many?" with count nouns.

### When to use the indefinite article *a* or *an*

Use *a* or *an* before a singular count noun when you do not specify which one.

➤ It's hot in here. Please open *a* window. [any of several windows]

Use *a* before words beginning with a consonant sound and *an* before words beginning with a vowel sound.

| | |
|---|---|
| a *user* | an *hour* from now |
| an *advertising* agency | a *happy* family |

Use *an* before a silent *h (hour)* and *a* before an *h* with a consonant sound (*happy*).

### When to use the definite article *the*

Use the definite article before nouns that refer to something that the reader already knows about or that is obvious from the context.

➤ Please open ^the^ window. [The reader knows which window.]

Use the definite article after the first time a noun is mentioned.

➤ Fill a large pot with *water*. Bring ^the^ water to a rolling boil.

Use the definite article for items of unique reference.

➤ The ^Women's Suffrage Movement achieved its goal in 1920.

Use the definite article with nouns that are described by restrictive phrases or clauses.

➤ The woman ~~A woman~~ *on the left* is the chairperson.

➤ The person ~~A person~~ *who began the consumer rights movement* was Ralph Nader.

### When not to use the definite article *the*

In general, do not use *the* with most singular proper nouns: names of people (Julio Romero); names or abbreviations for businesses and universities (International Business Machines, NYU); languages or academic subjects (Spanish, history); most holidays or religious occasions (Labor Day, Passover); names of most streets, parks, cities, and states (Main Street, Yellowstone National Park, Salt Lake City, Arizona); names of most countries and continents (Brazil, Africa); and names of bays, single lakes, single mountains, and islands (Hudson Bay, Lake Erie, Mount Etna, Catalina).

Notable exceptions include large geographic regions, oceans, seas, gulfs, deserts, mountain ranges, peninsulas, canals, and rivers (the Middle East, the Atlantic, the North Sea, the Gulf of Mexico, the Mohave Desert, the Himalayas, the Yucatan Peninsula, the Suez Canal, the Mississippi River).

## 29b Helping verbs and main verbs

Verb constructions—main verbs (MV) and helping verbs (HV), especially modals—can be troublesome for nonnative speakers. Modals express probability, necessity or obligation, or ability.

➤ He *might* (HV) *wash* (MV) the dishes. [probability]

➤ He *should* (HV) *wash* (MV) the dishes. [obligation]

➤ He *can* (HV) *wash* (MV) the dishes. [ability]

Other helping verbs include the forms of *have, be, do, used to,* and *ought to.*

➤ He *is washing* the dishes. (HV: *is*; MV: *washing*)

➤ He *did wash* the dishes. (HV: *did*; MV: *wash*)

➤ He *used to wash* the dishes. (HV: *used to*; MV: *wash*)

### Forms of verbs

All verbs have five forms, with the exception of *be,* which has eight forms. The first three of the five verb forms are called the *principal parts of a verb.*

| | | |
|---|---|---|
| 1. BASE FORM OR INFINITIVE | *(to) talk/write* | *(to) be* |
| 2. PAST TENSE | I *talked/wrote* | *was, were* |
| 3. PAST PARTICIPLE | I *have talked/written* | *been* |
| 4. PRESENT TENSE | I *talk/write*<br>You *talk/write*<br>He/she/it *talks/writes* | *am*<br>*are*<br>*is* |
| 5. PRESENT PARTICIPLE | I *am talking/writing* | *being* |

### Modal + base form

The modals *can, could, may, might, must, should,* and *will* can be used with the base form of the verb to write about the present or the future.

➤ Jeff can ~~swimming~~ swim very well. He should ~~competes~~ compete.

### *Do, does,* or *did* + base form

Use the base form of the main verb after all forms of the helping verb *do.*

➤ Ahmed does not invest~~s~~ in the stock market.

### Form of *have* + past participle

The perfect tenses are created by combining *have, has,* or *had* with the past participle of the main verb (usually ending in *-ed, -d, -en, -n,* or *-t*).

➤ Luis and Lupe *have* ~~*live*~~ lived here for ten years.

➤ We already *had* ~~*eat*~~ eaten dinner when the Smiths arrived.

**Form of *be* + present participle**

The progressive tenses consist of a form of the helping verb *be* (*am, is, are, was, were, being, be,* or *been*) plus the present participle of the main verb (the *-ing* form).

➤ I *am taking* the bus this week while the mechanic fixes my car.

➤ Sheila *was washing* her hair when the doorbell rang.

Certain verbs are *typically* not used in the progressive tenses (though there are exceptions). These verbs express states of being, emotions, or sense perceptions—for example, *appear, be, become, have, seem, believe, disagree, dislike, hate, imagine, intend, know, like, pity, prefer, realize, suppose, think, understand, want, wish, wonder, feel, hear, see, smell,* and *taste.*

➤ Cheng ~~was seeming~~ seemed sick all morning.

➤ The milk ~~is smelling~~ smells sour.

**Form of *be* + past participle**

The passive voice consists of a form of *be* (*am, is, are, was, were, being, be,* or *been*) plus the past participle of the main verb (usually ending in *-ed, -d, -en, -n,* or *-t*).

➤ Many electronic parts are ~~manufacture~~ manufactured in Southeast Asia.

➤ In my family, clothing was ~~wore~~ worn by all three children and then ~~gave~~ given to the church.

Use only transitive verbs in the passive voice because they always take a direct object. You cannot form the passive voice with intransitive verbs because they have no direct object. If you are not sure whether a verb is transitive, check your dictionary.

**Two-word verbs**

Two-word verbs consist of a main verb plus a particle. Note [illegible] particle changes the meaning of a verb completely.

➤ let down [disappoint]

➤ let out [free]

These combinations of verb + particle are idiomatic; that is, the separate meanings of the verb and the particle do not predict the meaning when they are combined into a two-word verb.

Like other verbs, two-word verbs may be either transitive or intransitive.

**TRANSITIVE** come across [find], tell off [rebuke]

**INTRANSITIVE** catch on [understand], crop up [occur]

Transitive two-word verbs are either inseparable or separable. The verb and particle of inseparable two-word verbs are never separated by the direct object (DO).

➤ Olga does not *take after* her older sister. [DO: her older sister]

However, the verb and particle of separable two-word verbs are sometimes separated by the direct object.

➤ We *turned* the offer *down*. [DO: the offer]

There is no way of telling which two-word verbs are inseparable and which are separable except by noting how a writer handles the object of these verbs. You must, therefore, note this distinction when you learn each new two-word verb.

**SEPARABLE**

| | |
|---|---|
| *fill out* an application/*fill* it *out* | (to complete) |
| *give up* smoking/*give* it *up* | (to quit) |
| *look up* some words/*look* them *up* | (to find, usually in a book) |
| *put out* a fire/*put* it *out* | (to extinguish) |
| *tear down* a house/*tear* one *down* | (to destroy or demolish) |
| *throw away* the paper/*throw* it *away* | (to discard) |
| *try on* some clothes/*try* some *on* | (to put on and examine) |
| *turn down* the TV/*turn* it *down* | (to reduce) |
| *turn up* the TV/*turn* it *up* | (to increase) |

INSEPARABLE

| | |
|---|---|
| *call on* a student/*call on* her | (to ask someone to speak) |
| *come across* an article/*come across* it | (to encounter something by chance) |
| *get over* a cold/*get over* it | (to recover) |
| *hear from* an old friend/*hear from* him | (to receive a communication) |
| *look after* his mother/*look after* her | (to take care of) |
| *look through* an encyclopedia/*look through* it | (to browse) |
| *pick on* the class clown/*pick on* someone | (to harass or bother) |
| *run into* a neighbor/*run into* her | (to encounter somebody by chance) |
| *take after* a grandparent/*take after* him or her | (to be, act, or look like) |

## 29c Prepositions and prepositional phrases

A preposition shows a relationship between objects or ideas in a sentence. The relationship can be in space (spatial), in time (temporal), or in manner.

SPATIAL Ravi's room is *next to* the candy machine.

TEMPORAL Dr. Bey has her office hours *before* class.

MANNER They eat french fries *with* lots of ketchup.

A prepositional phrase contains a preposition followed by a noun or noun phrase.

### Using the correct preposition

Not all languages have prepositions. Languages that do have them usually have fewer prepositions than English does, and the meanings of prepositions in other languages often do not correspond to the meanings of English prepositions. Moreover, most English prepositions have more than one meaning. (He left *at* 4:00. I live *at* 28 DeForest Heights.)

When you are unsure of which preposition to use, consult your dictionary, or ask a native speaker.

➤ Yesterday I worked all day ~~to~~ at/in the library. [*To* is used to show direction with motion verbs, as in the sentence *I go to school every day. Work* does not show motion.]

➤ Susan and Frank live ~~at~~ in Chicago. [*At* is used with exact addresses, as in the sentence *Susan and Frank live at 55 Summit Street. In* is used with the names of cities and states. Chicago is a city.]

## 29d Adjectives

*Adjectives* show the quality or number of a noun or noun substitute.

### Word order with multiple adjectives

Use idiomatic word order when you modify a noun with more than one adjective.

➤ ~~an oval small~~ a small oval gem five ~~red brick tall~~ tall red brick buildings

The following word order is typically used when more than one adjective precedes a noun.

1. Article, possessive adjective, or quantifier: *a/an, the, my, people's, several, some.*
2. Number or comparative or superlative form of the adjective: *two, first, slower, slowest.*
3. Evaluative adjective that can be modified by *very*: *beautiful, committed, delicious.*
4. Size: *big, small, long, short.*
5. Shape: *round, square, triangular.*
6. Age: *modern, nineteenth-century, ancient, new, old.*
7. Color: *red, green, mauve.*
8. Nationality: *German, Chinese, Mexican.*
9. Religion: *Protestant, Jewish, Muslim, Buddhist.*
10. Material: *wood, brick, gold.*
11. Noun used as an adjective: *student* lounge, *music* program, *dining room* furniture.

## 29e Omitted verbs, subjects, or expletives

English requires a verb in each clause or sentence, even when the meaning is clear without it. Sentences need *be* to link a subject with a prepositional phrase or a predicate adjective.

➤ The conference ^is in New York this year.

➤ New Yorkers ^are very friendly.

Every sentence or independent clause has, as a rule, its own subject.

➤ The weather, which started out cloudy, ~~it~~ improved later in the day. [Even though the subject, *weather,* is separated from the verb, *improved,* by a descriptive clause, it should not be repeated.]

In most English sentences, the subject is first, and the verb and other parts of the predicate follow it. However, a subject may be postponed until later in the sentence. In this case, an expletive (*there, here,* or *it*) at the beginning of the sentence substitutes for the postponed subject.

Expletives are used in the following ways:

- *There* is a common expletive used to indicate that a particular phenomenon exists.

  ^There is ~~Is~~ a large Hmong community in Minneapolis. [The subject, *a large Hmong community,* is placed after the verb.]

- The expletive *here* usually introduces information.

  *Here* are a few ways to be an effective language learner.

- The pronoun *it* often functions as an expletive when the deferred subject is a noun clause or phrase.

  *It* is tragic to watch our tropical rain forests disappear. [The subject is the entire noun phrase.]

  *It* is not enough that we have good intentions. [The subject is the noun clause *that we have good intentions.*]

# *Punctuation*

# 30 The Comma

Commas help communicate meaning by eliminating possible misreadings. Consider the following sentence, which is missing internal punctuation, and three revisions of it:

➤ After telephoning Paul Lee James went to the library.

➤ After telephoning**,** Paul Lee James went to the library.

➤ After telephoning Paul**,** Lee James went to the library.

➤ After telephoning Paul Lee**,** James went to the library.

## 30a Commas with independent clauses joined by a coordinating conjunction

Place a comma before a coordinating conjunction (*and, but, for, nor, or, so,* or *yet*) that joins two or more independent clauses.

➤ We visited the Museum of Fine Arts in Boston^, and we both saw the Winslow Homer exhibit.

➤ We knew the Natalie Cole concert would be popular^, so we bought our tickets early.

EXCEPTION It is acceptable to omit the comma, especially before *and* or *or,* when the main clauses in a compound sentence are short and there is no possibility of misreading.

➤ Show me a hero and I will write you a tragedy.
—F. Scott Fitzgerald

But in your own prose, you'll never go wrong by inserting a comma between independent clauses joined by a coordinating conjunction. (See also page 97.)

## 30b Commas with introductory word or word group

Generally, use a comma to separate an introductory word or word group from the main part of your sentence. With *short* introductory word groups, you sometimes can omit the comma, but you are almost never wrong to insert a comma (exceptions are marked by an asterisk). When misreading might occur, you *must* insert the comma.

| EXAMPLE | RULE |
|---|---|
| When electric power fails, hospitals rely on emergency generators. When we got in, the canoe lurched. | Commas to set off introductory clause: without commas, readers might overlook the important pause between *fails* and *hospitals* and between *in* and *the canoe.* |
| *Since the Soviet Union collapsed Russians have struggled economically. | EXCEPTION Do not use a comma after an introductory clause beginning with *since* when *since* expresses time. |
| In his book on Nepal, Trank tells us how he came to live there. | Comma to set off introductory prepositional phrases: without comma, readers might think *Nepal Trank* is a compound noun. |
| Baking, the cookies released a wonderful aroma. Working out every day, Maria made the team. | Comma to set off participle and participial phrase: without a comma in the first sentence, readers might think *Baking the cookies* is the subject of the sentence. |
| To catch the ring, reach out earlier. *To catch the ring is my goal. | Comma to set off infinitive phrase. EXCEPTION Do not set off infinitive phrase that is the subject of the sentence. |
| Everything considered, we chose a perfect day for the picnic. | Comma to set off absolute phrase |

## 30c Commas with nonrestrictive elements

Adjective phrases, adjective clauses, and appositives are either restrictive or nonrestrictive. A restrictive element contains essential information that limits—or restricts—the meaning of the noun or pronoun it describes. Do not set off a restrictive element with commas. (See also pages 92–93.)

➤ The woman *wearing the green print dress* plays basketball.

➤ The teachers *who had gone on strike* lost their jobs.

➤ Shakespeare's heroine *Kate* is my favorite.

When an adjective phrase, adjective clause, or appositive isn't critical to identifying the noun or pronoun it modifies, it is nonrestrictive; you can omit the element without changing the meaning of the sentence. Therefore, you should set it off with commas. (See also page 93.)

➤ The woman**,** *looking for a new apartment***,** plays basketball.

➤ The teachers**,** *who averaged ten years of service***,** lost their jobs.

➤ In Shakespeare's *The Taming of the Shrew***,** the female lead**,** *Kate***,** is my favorite.

## 30d Commas to separate items in a series

➤ Rob bought his calculator**,** textbooks**,** and diskettes at the campus bookstore.

➤ The horses trotted down the hill**,** across the meadow**,** and into the barn.

Although it is not wrong to omit the comma before the final conjunction in a series, most writers use the final comma to avoid any risk of confusion or misreading.

## 30e Commas to separate coordinate adjectives

The adjectives in the following example are coordinate adjectives because each one modifies a noun separately.

➤ The *dark***,** *muddy***,** *smelly* swamp was infested with mosquitoes.

Coordinate adjectives take commas between them. You can join coordinate adjectives with *and* instead of with commas (*dark* and *muddy* and *smelly*); you can also change the order of coordinate adjectives without altering the meaning of the sentence (The *muddy*, *dark*, *smelly* swamp was infested with mosquitoes).

The adjectives in the following example are cumulative adjectives. Each adjective modifies the adjective that follows it as well as a noun or pronoun.

➤ Angela sewed *two large denim* patches on the knees of her jeans.

The adjectives do *not* take commas or *and* between them because *denim* modifies *patches, large* modifies *denim patches,* and *two* modifies *large denim patches.* You can't change the order of these adjectives without getting a non-idiomatic sentence.

## 30f Commas with parenthetical and transitional expressions

Use commas to set off a parenthetical expression, which you can insert in a sentence to qualify, explain, or give your point of view.

➤ In some kinds of writing**,** especially formal writing**,** contractions are frowned upon.

Use commas to set off a transitional expression, which shows logical connections.

➤ As a matter of fact**,** the expedition will be extremely dangerous.

➤ A turkey sandwich**,** for example**,** is a healthy lunch.

➤ You should**,** therefore**,** keep your credit cards in a safe place.

## 30g Commas with contrasted elements

Expressions that indicate a contrast or contradiction usually begin with *not, nor, but,* or *unlike.*

➤ The steelworkers, unlike management, wanted a shorter workweek.

## 30h Commas with direct quotations

➤ "Push the red button," she explained, "to record your message."

## 30i Commas with mild interjections, words of direct address, the words *yes* and *no*, and interrogative tags

➤ Well, I think a compromise is in order.

➤ Your sculpture of the wild horse is excellent, Susan.

➤ Yes, we'll meet you at the gym tomorrow.

➤ The university has a policy about cars on campus, doesn't it?

## 30j Other conventional uses of commas

TITLES

➤ Priscilla Frank, DVM, has a practice in Miami.

DATES

➤ On Friday, February 28, 1999, Sarah turned twenty-nine years old.

EXCEPTIONS Omit the commas when the date is inverted (28 February 2002) or when only the month and year are given (February 2002).

ADDRESSES

➤ Angela's new address is 1566 Fifield Street, St. Paul, MN 55100.

PLACE NAMES

➤ British Airway's flight 404 leaves Atlanta, Georgia, on Sunday morning and arrives in Paris, France, that evening.

NUMBERS

➤ CBS received 12,500 letters after the Arafat interview.

## 30k Misuses of the comma

Do not use a comma in the following situations:

**BETWEEN A SUBJECT AND ITS VERB**

➤ Candidates who fail to file timely financial reports, are subject to fines.

**BETWEEN AN ADJECTIVE AND A NOUN**

➤ Affirmative action is a sensitive, difficult, issue.

**BETWEEN COMPOUND ELEMENTS THAT ARE NOT INDEPENDENT CLAUSES**

➤ Paul picked up his suit at the cleaners, and had a prescription filled.

➤ To sell, or not to sell is the question stock speculators face daily.

**EXCEPTION** You may need a comma between parts of a compound predicate if you think misreading will occur without it.

**AFTER THE WORDS *SUCH AS* OR *LIKE***

➤ She reads popular magazines such as, *Time*, *Money*, and *People*.

**BEFORE *THAN***

➤ Coming in third is definitely more satisfying for me, than placing any further back.

**AFTER A COORDINATING CONJUNCTION**

➤ The Yankees won the pennant last year but, not this year.

**BEFORE A PARENTHESIS**

➤ Use the current name of the publisher, (e.g., Harcourt Brace), not the name at the time of publication.

**WITH AN INDIRECT QUOTATION**

➤ When I got through voice mail, the secretary said, that she would take a message.

**WITH A QUESTION MARK OR AN EXCLAMATION POINT**

➤ "I love you!," she shouted across Grand Central Station.

# 31 The Semicolon

A semicolon joins independent clauses not joined by a coordinating conjunction and not separated by a period. A semicolon also separates phrases and clauses containing other punctuation.

COMMA TOO WEAK

➤ Gazpacho is only the first course**,** the complete dinner is five courses.

SEMICOLON APPROPRIATE

➤ Gazpacho is only the first course**;** the complete dinner is five courses.

PERIOD APPROPRIATE

➤ Gazpacho is only the first course**.** The complete dinner is five courses.

Generally, readers are confused by a comma that is too weak to hold two independent clauses together. Therefore, the following rules focus on replacing inadequate commas with semicolons.

## 31a Semicolon between closely related independent clauses not joined by a coordinating conjunction

A semicolon joining two independent clauses indicates a strong connection between the content of the two clauses.

➤ Anatomy is not destiny**;** it is simply anatomy.
—Elizabeth Janeway

Using only a comma to join two independent clauses creates the error commonly known as the comma splice (see pages 80–81).

➤ The new cars are really disappointing~~,~~;they all look alike.

In general, when a coordinating conjunction joins two independent clauses, use a comma—not a semicolon—to punctuate the resulting compound sentence (see page 92). If, however, at least one of the independent clauses contains its own comma, then you may elect to use a semicolon.

➤ We knew the concerts with Natalie Cole, Britney Spears, and Paul Simon would be popular; so we bought all our tickets early.

## 31b Semicolon between independent clauses when a conjunctive adverb or transitional expression introduces the second independent clause

➤ There are no telephones or radios at Rio Caliente; *however,* television is available by satellite hookup.

➤ Cal-Edison recently instituted several new environmental policies; *for example,* it now offers customers a number of incentives to do home-energy audits.

Here is a list of words that can function as conjunctive adverbs:

| | | |
|---|---|---|
| accordingly | however | similarly |
| also | incidentally | specifically |
| anyway | indeed | still |
| besides | instead | then |
| certainly | likewise | thereafter |
| consequently | meanwhile | therefore |
| conversely | moreover | thus |
| finally | nevertheless | undoubtedly |
| furthermore | now | |
| hence | otherwise | |

Here is a list of words that can function as transitional expressions:

| | | |
|---|---|---|
| afterward | here | in other words |
| below | in addition | of course |
| elsewhere | in conclusion | on the other hand |
| for instance | in fact | |

## 31c Semicolon between items in a series containing their own commas

In the following example, the semicolons help the reader distinguish major sentence elements and thus more quickly and accurately understand the writer's intended meaning.

➤ The works in the American novel course include Hawthorne's *The Scarlet Letter,* with the love triangle of Hester Prynne, Arthur Dimmesdale, and Roger Chillingworth; Melville's *Moby-Dick,* with the monomaniacal Captain Ahab; and Chopin's *The Awakening,* with the heroic struggle of Edna Pontellier.

## 31d Misuses of the semicolon

The sentence elements joined by a semicolon must always be of equal grammatical rank. Do not use semicolons to join an independent clause and a phrase, an independent clause and a subordinate clause, or an independent clause and a list.

➤ While doing research in Arizona, Bill visited the Sonora Desert Museum.

➤ Polly took a short nap, which she deserved after baby-sitting all day.

➤ The grocery store carried difficult-to-find fruits and vegetables: Asian pears, star fruit, anise, and bok choy.

# 32 The Colon

Use the colon after an independent clause to direct your reader's attention forward to the word(s), phrase(s), or clause(s) that follow it.

## 32a Colon to introduce an explanation

➤ I am a man who at the precocious age of thirty-five experienced an astonishing revelation: it is better to be a success than a failure. —Norman Podhoretz

## 32b Colon to introduce a series

➤ The following constitutes a basic personal reference library**:** a college dictionary, an almanac, a thesaurus, an atlas, and an English handbook.

➤ Camp Evergreen offers several waterfront activities**:** swimming, canoeing, fishing, windsurfing, and water skiing.

## 32c Colon to introduce an appositive

➤ In the Mideast, public attention focused on one thing**:** peace.

## 32d Colon to introduce a direct quotation

➤ I'll always remember the Chinese proverb I learned in the ninth grade**:** "I hear and I forget. I see and I remember. I do and I understand."

## 32e Colon to mark conventional separations

SALUTATION AND BODY OF A BUSINESS LETTER Dear Manager**:**

TITLE AND SUBTITLE The *Army of the Potomac***:** *A Stillness at Appomattox*

DATE AND PAGE IN BIBLIOGRAPHIC ENTRIES *Time* 27 Oct. 1997**:** 62–72.

PLACE AND PUBLISHER IN BIBLIOGRAPHIC ENTRIES New York**:** Longman

CHAPTER AND VERSE IN A BIBLICAL CITATION Ruth 2**:**3
[Note: MLA style recommends a period between chapter and verse (Ruth 2.3).]

## 32f Misuses of the colon

Do not use a colon to separate a verb and its object, a verb and its complement, or a preposition and its object.

➤ We need to buy~~:~~ lettuce, peppers, onions, cucumbers, and carrots.

➤ The candidates for Student Association president are: Holden and Dwire.

➤ They sold a selection of: baskets, alpaca sweaters, and silver jewelry.

# 33 The Apostrophe

An apostrophe shows possession; indicates omitted letters and numerals; and forms the plurals of letters, numbers, abbreviations, and words used as words.

## 33a Apostrophe to mark the possessive case

Guidelines for the possessive case relate to considerations of final sounds and letters.

### Possessive of singular and plural nouns and indefinite pronouns that do not end in *s* or with an *s* or *z* sound

Add an *'s* to form the possessive.

| | |
|---|---|
| Joanne's house | that man's car |
| anybody's guess | those men's cars |

### Possessive of singular nouns ending in *s* or with an *s* or *z* sound

Add an *'s* to form the possessive.

| | |
|---|---|
| Keats's poetry | Konrad Lorenz's research |
| the boss's car | the prince's cloak |

### Possessive of plural nouns ending in *s* or with an *s* or *z* sound

Add only an apostrophe to form the possessive.

| | |
|---|---|
| the boys' coats | the Smiths' house |
| the dogs' bones | the golfers' tour |

#### Possessive of compound nouns

Add *'s* to only the last word to form the possessive.

➤ my mother-in-law's clock

➤ the runner-up's prize

#### Joint possession

To show joint ownership, make only the last noun possessive.

➤ Tom and Jennifer's house [Tom and Jennifer own the house together.]

To show individual ownership, make both nouns possessive.

➤ Tom's and Jennifer's philosophies. [Tom and Jennifer have different philosophies.]

## 33b Apostrophe to indicate contractions

The apostrophe can replace one or more omitted letters or numbers.

➤ I'm [I am]

➤ fish 'n' chips [fish and chips]

➤ the earthquake of '94 [1994]

## 33c Apostrophe and *s* to pluralize letters, numbers, abbreviations, and words cited as words

➤ Your *m*'s look like *w*'s.

➤ Count to 99 by 3's.

➤ Most banks have ATM's today.

➤ I have heard enough of your *no*'s for a while.

A letter, number, or word used as a word is italicized, but the *s* following the apostrophe is not.

EXCEPTION The apostrophe is generally not used to form the plural of a decade.

The 1970s had much in common with the 1920s.

**NOTE** MLA style calls for no apostrophe with plurals of numbers and abbreviations: 3s (or threes), ATMs.

## 33d Misuses of the apostrophe

Do not use an apostrophe in the following situations:

**WITH POSSESSIVE PRONOUNS**

➤ The pizza with anchovies is all your's.

**WITH POSSESSIVE PRONOUNS THAT SOUND LIKE CONTRACTIONS**

➤ its it's [it is]

➤ your you're [you are]

➤ whose who's [who is, who has]

➤ their they're [they are]

**IN REGULAR PLURAL NOUNS**

➤ Some cyclist's ride without helmets.

**WITH NAMES OF SOME ORGANIZATIONS**

Follow the usage preferred by organizations. Check a dictionary or publications by the organization.

➤ Veterans Administration

➤ Women's Health Center

# 34 Quotation Marks

Quotation marks are used in pairs to signal the start and the end of a direct quotation, to enclose titles of short works, and to indicate words used as words.

## 34a Quotation marks with direct quotations

Direct quotations are written or spoken words repeated verbatim and must be in quotation marks. Integrate quotations of four or fewer typed lines of prose or three or fewer lines of poetry into the text of your paper.

➤ It was Kahlil Gibran who said, “Work is love made visible.”

➤ Walt Whitman boldly states, “I too am not a bit tamed, I too am untranslatable, / I sound my barbaric yawp over the roofs of the world.” [The slash (/) marks the end of a line of poetry. See page 114.]

Block off prose quotations that are longer than four typed lines and poetry quotations that are more than three lines. Do not use quotation marks with blocked-off quotations. (See pages 172–173.)

Never use quotation marks with indirect quotations, which report what someone has said instead of citing that person’s exact words.

➤ Hector told me he couldn’t remember where he’d left his biology book.

Set off speech tags such as *she said* and *I replied* with commas. When you use quotation marks to indicate dialogue, the directly quoted speech of two or more people, signal a change of speaker by beginning a new paragraph.

Use single quotation marks (‘ ’) to enclose a quotation within a quotation.

➤ Professor Cohen said, “I will always remember Ellery Sedwick’s advice, ‘Autobiographies ought to begin with the second chapter.’ ”

## 34b Quotation marks to indicate the titles of short works

Use quotation marks around the titles of articles, essays, short stories, poems, chapters of a book, songs, and episodes of radio or television programs.

➤ Elton John’s song “Candle in the Wind” has sold millions of copies.

Use italics or underline the titles of longer works—including the names of newspapers and magazines and the titles of books, plays, films, and television or radio series. (See pages 123–124.)

## 34c Quotation marks to indicate words used as words

Editors still prefer italics or underlining to quotation marks for words used as words (see page 124), but quotation marks are acceptable.

➤ I finally understand the difference between “mean” and “median.”

## 34d Other marks of punctuation with quotation marks

### Commas and periods inside quotation marks

➤ “Now**,**” the conductor said**,** “let’s add the violins**.**”

When using MLA parenthetical citations, place the parenthetical citation after the concluding quotation mark and before the period. (See pages 178–184.)

➤ According to Davis**,** “One of the most potent elements in body language is eye behavior” (27)**.**

### Colons and semicolons outside quotation marks

➤ Mount Rushmore is a striking memorial to four presidential “giants”**:** Washington, Jefferson, Lincoln, and Theodore Roosevelt.

➤ All domestic flights are now designated “no smoking”**;** the rule is clearly posted and strictly policed.

### Question marks and exclamation points with quotation marks

Place question marks and exclamation points inside quotation marks when they apply to the quoted material; place question marks and exclamation points outside quotation marks when they apply to the whole sentence.

➤ Harriet said, “Will you meet me at Bloomingdale’s**?**”

➤ Stop whispering “The Star Spangled Banner”**!** Sing it out proudly**!**

## 34e Misuses of quotation marks

Use quotation marks judiciously, or you will diminish their intended effect. Do not use quotation marks in the following situations:

AROUND NICKNAMES, SLANG, TECHNICAL TERMS, OR AFTER THE WORD *SO-CALLED*

➤ The kids all called her "Bunny."

➤ In weaving class we learned about the "warp" and the "weft."

WITH INDIRECT QUOTATIONS

➤ Coach Barnett told his players "to be on time for practice."

Finally, do not use quotation marks around the titles of your papers.

# 35 Other Marks

## 35a Periods with most sentences and some abbreviations

### Most sentences

Use a period to mark the end of straightforward statements and commands but not the end of direct questions and exclamations.

STATEMENT

➤ Last month the Ford dealership sold forty-seven new cars.

COMMAND

➤ Put your dishes in the sink when you finish eating.

If your sentence tells of a question that was asked as opposed to asking the question directly, end that sentence with a period.

INDIRECT QUESTION

➤ Mary asked when Annie Dillard had written *Pilgrim at Tinker Creek*.

Standard abbreviations

Use a period after standard abbreviations.

| | | |
|---|---|---|
| Rev. | St. | Capt. |
| etc. | Nov. | a.m. |
| Inc. | Dr. | e.g. |

When an abbreviation falls at the end of a sentence, use only one period.

➤ Mary told her friends she had accepted a job with Desktop Graphics, Inc.

Do not use a period for abbreviations that are usually pronounced as words (UNESCO) or sounded as letters (FBI).

| | | |
|---|---|---|
| ESP | TNT | NBA |
| TV | KO | IRS |
| CARE | NOW | NAFTA |

Do not use periods with the United States Postal Service's abbreviations for states.

AR CA FL MI ND TX WA WY

NOTE Because usage varies regarding different kinds of abbreviations, it is always a good idea to consult your college dictionary or a style manual when you are uncertain about a particular one.

## 35b Question mark with all direct questions

➤ When was Beethoven born?

When a direct question occurs within a declarative sentence, use a question mark at the end of the question and a period at the end of the sentence.

➤ "Where do you recommend we go to dinner tonight?" Gary inquired.

When you write a polite request in the form of a question, you may end it with a question mark, though some writers prefer a period.

➤ Will you please lock the door when you leave tonight**?**

## 35c Exclamation point with strong statements

Use an exclamation point at the end of a sentence that deserves special emphasis or expresses strong emotion, but use this mark of punctuation sparingly, or it will lose its impact in your writing.

➤ "Ouch**!**" he screamed.

➤ The results of the survey are alarming**!**

## 35d Dash for interruptions and shifts

Use a dash or a pair of dashes to emphasize part of a sentence or to set it off for clarity. Dashes lose their intended effect, however, when used indiscriminately as substitutes for commas, semicolons, or periods.

On a computer, you can specify an "em" dash from a symbols list, or you can use two unspaced hyphens (--) to substitute for a dash (—); do not leave a space before or after the hyphens or the dash. (See pages 125–127 for other uses of the hyphen.)

### Dashes for emphasis

Use dashes to emphasize informal explanations and afterthoughts.

➤ Ads for bank credit cards and department stores refer to "convenient terms"—meaning 18 percent annual interest rates payable at the convenience of the creditor.

—*Time* magazine

➤ Our civilization is decadent, and our language—so the argument runs—must inevitably share in the general collapse.

—George Orwell

### Dashes for appositives

A nonrestrictive appositive is set off with commas (see page 94), but if the appositive itself contains commas, use dashes to clarify how this one part of the sentence fits into the whole.

➤ All our preparations for spring break—choosing the best vacation package, shopping for new clothes, getting to the tanning salon—made it hard to study for midterms.

### Dashes for reversals

Use dashes to signal a dramatic reversal of thought or tone.

➤ He was a terrific father—when he was out of town.

## 35e Parentheses with nonessential information

Parentheses indicate helpful but not essential information. Avoid overusing parentheses, or it will appear that your paper is cluttered with nonessential information. As you write and revise, check to see whether you can elevate some parenthetical elements to part of the text by using commas or dashes. Perhaps some details are so nonessential they can be dropped altogether.

➤ Anita Jennings, ~~(~~president of Harris Bank~~)~~, thinks treasury notes are a good investment.

➤ Each year twenty-five thousand — ~~(~~some studies claim as many as fifty thousand~~)~~ — Americans die in alcohol-related accidents.

➤ Write your answers neatly ~~(and clearly)~~ in the left column.

### Parentheses to enclose supplementary or interpretive information, asides, and afterthoughts

➤ The average homemaker works just under one hundred hours a week at a variety of jobs (purchasing agent, cook, cleaner, economist, chauffeur, etc.) with no annual salary and little recognition.

➤ For almost ten years (and that was ten years too long), American soldiers fought in Vietnam.

### Parentheses to enclose numbers or letters used to mark a series in a sentence

➤ Bacteria are divided into three groups based on shape: (1) spheres, (2) rods, and (3) spirals.

## 35f Brackets with changes in quoted material

### Brackets to add clarification

Use brackets to add or substitute a clarifying comment in quoted material. In the following example, by using brackets to insert the kinds of problems Carson is referring to, the writer avoids having to quote an extended passage.

➤ The environmentalist Rachel Carson believes that "problems [with controlling insects] arose with the intensification of agriculture—the devotion of immense acreages to a single crop."

### Brackets with an ellipsis

Use brackets around an ellipsis you have inserted within a quotation to indicate omitted material. (For a discussion of brackets with ellipses, together with examples, see pages 112–113.)

### Brackets to indicate error in original

The Latin word *sic*, meaning "thus," appears within brackets in a direct quotation to indicate that you recognize an error in the original and are letting it stand.

➤ One film critic believed that "after there [sic] first encounter with the Wizard, the main characters were more determined."

## 35g Ellipsis mark to indicate omissions from quotations

The ellipsis mark is three equally spaced periods ( . . . ) indicating that the writer has omitted words from quoted material. An ellipsis mark tells the reader that you have omitted the part of a quotation that is not pertinent to your point; at the same time, it shows that you are not misrepresenting the person being quoted. To distinguish your ellipses from those that may appear in the original text, place brackets around the ellipsis marks that you add.

➤ The secretary of transportation said "drivers **[. . .]** must learn to use seat belts because seat belts save lives."

When inserting an ellipsis, leave a space before the first bracket, a space before the second and third periods of the ellipsis but no space before the first or after the third, and then leave a space after the last bracket.

### Ellipsis mark to signal omission of one word or whole sentences from a quoted passage

If omitted words within the original passage follow some form of internal or end punctuation, retain that punctuation, and then insert the ellipsis mark.

> "Youth is impulsive. When our young men grow angry **[. . .]** and disfigure their faces with black paint, it denotes that their hearts are black**, [. . .]** and our old men and old women are unable to restrain them**. [. . .]** Thus it was when the white men first began to push our forefathers further westward. But let us hope that the hostilities between us may never return."
>
> —Chief Seattle, "My People"

OMISSION OF WORDS WITHIN SENTENCE

OMISSION OF WORDS WITHIN SENTENCE—RETAINED PUNCTUATION

OMISSION OF WHOLE SENTENCE

Do not use an ellipsis mark at the beginning of a quotation. Do not use an ellipsis mark at the end of a quotation unless you have deleted words from the end of the final sentence. In that case, leave a space before the first bracket and immediately follow the last bracket with the sentence period and the closing quotation mark:

➤ "At the beginning of the Olivier *Hamlet* film, we are told by Olivier's disembodied voice that this is the story of a man who could not make up his mind. Plays do not normally do this sort of self-foretelling **[...].**"
—Bert O. States, *Great Reckonings in Little Rooms*

When making an in-text parenthetical citation after the ellipsis at the end of your sentence, leave a space before the first bracket and immediately follow the last bracket with the closing quotation mark, a space, the parenthetical citation, and the sentence period.

➤ "Plays do not normally do this sort of self-foretelling **[...]**" (35)**.**

**To indicate the omission of a full line or more of poetry**

Use a single line of spaced periods within brackets.

I shall be telling this with a sigh
**[. . . . . . . . . . . . . . . . . . . . . . . . . . . . . . .]**
Two roads diverged in a wood, and I—
I took the one less traveled by,
And that has made all the difference.
—Robert Frost, "The Road Not Taken"

## 35h Slash with alternative words and between lines of poetry

Use a slash to indicate options or alternative words. Do not leave a space either before or after the slash when separating options.

➤ Marty took physics pass**/**fail.

Use a slash to mark off individual lines of poetry when you run two or three lines into your text. Leave one space before and after the slash when quoting poetry. When quoting four or more lines of poetry, set them off from the text as a block.

➤ Emily Dickinson captures the paradox of success in the opening lines: "Success is counted sweetest / By those who ne'er succeed."

# *Mechanics*

# 36 Capital Letters

The rules for capitalization are quite clear-cut; even the cases involving optional capitalization are clear. If you have a question about a particular word, consult your college dictionary.

## 36a Proper nouns and proper adjectives

NAMES OF PEOPLE AND THINGS

| | |
|---|---|
| Bill Cosby | William Least Heat-Moon |
| Calamity Jane | Halley's Comet |
| *Titanic* | *Apollo 16* |

Do not capitalize the names of academic subjects, except languages.

➤ Josie is taking chemistry, math, philosophy, German, and English.

Do capitalize, however, the names of specific courses.

➤ New students must take History 271 as well as Race and Culture.

PLACE NAMES

| | |
|---|---|
| Beijing | the Lone Star State |
| the Far East | the Nile River |

Do not capitalize points of the compass unless you are referring to specific regions.

➤ Jake went east for the holidays.

➤ Jake drove to the Northeast to see the colorful leaves.

RACES, NATIONALITIES, AND LANGUAGES

| | | |
|---|---|---|
| Native American | Latino | Japanese |
| African-American | (BUT black AND white) | |

NAMES OF TRADEMARKS

| | | |
|---|---|---|
| Levi's | Kleenex | Alka-Seltzer |
| WordPerfect | Hotbot | QuarkXPress |

Follow company and product style for innovative use of capitals in names. Do not capitalize nouns that follow brand names: *Scotch tape, Dodge truck.*

**NAMES OF HISTORICAL EVENTS, PERIODS, DOCUMENTS, AND TREATIES**

| | |
|---|---|
| French Revolution | World War II |
| Treaty of Versailles | the Renaissance |

Do not capitalize recently ascribed names of political, cultural, or scientific periods: *cold war, civil rights movement, space age.* Do not capitalize centuries: *the thirteenth century.*

**NAMES OF ORGANIZATIONS, CORPORATIONS, INSTITUTIONS, AND GOVERNMENT AGENCIES AND COURTS**

| | |
|---|---|
| Salvation Army | Phi Beta Kappa |
| Howard University | Citibank |
| Ford Foundation | Digital |
| United States Court of Appeals | Democratic Party |

When a common noun is part of a name, capitalize it.

| | |
|---|---|
| Xerox Corporation | Ohio River |
| Canadian Pacific Railroad | Rocky Mountains |

With plurals, however, do not capitalize common nouns: *United and Continental airlines.*

Do not capitalize words such as *a, an,* or *the* when used with names of organizations, corporations, and so on.

➤ For the New York Yankees, the 1980s were one long slump.

**NAMES OF DAYS, MONTHS, AND HOLIDAYS**

| | | |
|---|---|---|
| Saturday | October | Labor Day |

But do not capitalize the names of the seasons: *summer.*

**NAMES OF RELIGIONS AND RELATED TERMS**

| | | |
|---|---|---|
| Catholicism | Book of Mormon | Ramadan |
| Buddhist | Bible | Passover |

**PROPER ADJECTIVES**

| | | |
|---|---|---|
| Shakespearean | Victorian | Leninist |
| Christian | Arabian | Israeli |

Proper adjectives are derived from proper nouns.

## 36b Abbreviations

Use all capital letters in the abbreviations of organizations and government agencies, call letters for radio and television stations, and acronyms (words formed from the initial letters of a name).

| | | | | |
|---|---|---|---|---|
| AFL-CIO | NCAA | IBM | BART | PC |
| KOST-FM | WCAX-TV | USMC | AIDS | CD-ROM |

## 36c Titles with names

Capitalize a person's title if it appears before his or her name, but not when it appears after.

| | |
|---|---|
| Senator Daniel Inouye | Daniel Inouye, a senator |
| Judge Mary Ruiz | Mary Ruiz, a judge |

Governor Christine Todd Whitman
Christine Todd Whitman, a governor

When you use titles of world figures alone, capitalization is optional.

➤ The President [OR president] spoke to the reporters.

## 36d First word of a sentence, deliberate sentence fragment, quoted sentence, or independent clause after a colon

Always capitalize the first word of a sentence or a deliberate sentence fragment. When quoting a sentence, capitalize the first word.

➤ In an article on nuclear winter, Carl Sagan states, "Except for fools and madmen, everyone knows that nuclear war would be an unprecedented human catastrophe."

Do not capitalize the first word of a quoted sentence when you are blending it into the main sentence.

➤ Barbara Myerhoff has written that "our species can be characterized as *Homo narrans,* humankind as storyteller."

If you interrupt a quoted sentence with some words of explanation, do not capitalize the first word of the quotation after the break.

➤ "Debunking but sympathetic," says *Publishers Weekly,* "this is an engrossing biography of Frank Lloyd Wright."

When an independent clause occurs after a colon, capitalizing the first word is optional.

➤ The pennant race is too close to call: the [OR The] Red Sox and the Yankees will battle until the last day of the season.

## 36e First words of lines of poetry

Capitalize the first word of every line unless the poet does otherwise.

> For the man who should loose me is dead,
> Fighting with the Duke in Flanders,
> In a pattern called a war.
> Christ! What are patterns for?
> —Amy Lowell, "Patterns"

> what if a much of a which of a wind
> gives the truth to summer's lie;
> —e. e. cummings, "what if a much of a which of a wind"

## 36f First and last words and all other important words in titles of works

Unless they are the first or last word of a title or subtitle, do not capitalize articles (*a, an, the*), coordinating conjunctions (*and, but, or, for, nor, so, yet*), or prepositions (*in, at, of, near, from,* and so on).

BOOK *Tuesdays with Morrie*

BOOK/SUBTITLE *A Passion for Excellence: The Leadership Difference*

PLAY *The Adding Machine*

SHORT STORY "The Story of an Hour"

POEM "The Love Song of J. Alfred Prufrock"

FILM *The Green Mile*

PAINTING OR SCULPTURE *Homage to the Square*

ARTICLE "What's Wrong with Black English?"

COMIC STRIP *Hagar the Horrible*

SONG "If I Didn't Have You"

COMPACT DISC OR CASSETTE TAPE *Cinema Serenade*

# 37 Abbreviations

Except for conventional uses, as illustrated here, avoid abbreviations in most academic writing. Spell out the names of holidays, names of days and months, addresses (street, state, country), names of academic subjects and subdivisions of books, units of measurement, and parts of a business's or institution's name (unless the abbreviation is part of the official name).

## 37a Abbreviations of titles before and after proper nouns

According to MLA, you should omit periods and spaces in some abbreviations, especially those consisting mostly or entirely of capital letters.

| TITLES BEFORE THE NAME | TITLES AFTER THE NAME |
|---|---|
| the Rev. D. H. Shaw | Gerald R. Brown, Sr. |
| Dr. Virginia Wu | Eleanor T. Bates, DD |
| Adm. Hyman Rickover | Michael Kaplan, DVM |
| the Hon. Langdon Crane | Elizabeth Luria, PhD |

Do not use title abbreviations without a person's complete name. Always write out the title when no name is given.

➤ My ~~dr.~~ doctor is always late.

## 37b *AD, BC, a.m., p.m., no.,* and *$*

Use conventional abbreviations only with specific years, times, numbers, or amounts.

| | |
|---|---|
| 32 BCE* [OR BC] | 1812 CE* [or AD 1812] |
| 10,000 BP** | 8 a.m. |
| 6:30 p.m. | no. 33 |
| $250 | |

Always use BC ("before Christ") after the date and AD ("anno Domini") before the date.

*Before Common Era and Common Era.
**Before the Present.

## 37c Familiar abbreviations

By convention, certain names for organizations, corporations, government agencies, states, and countries are abbreviated with capital letters and no periods. Here are some of the most common:

| | | | |
|---|---|---|---|
| NAACP | AP | IRS | NY |
| NCAA | IBM | UK | TN |
| NATO | UPS | USA | ID |

For unfamiliar abbreviations in an essay, write out the complete name the first time you use it with the abbreviation immediately following in parentheses: *Modern Language Association* (*MLA*). Then use the abbreviation in the rest of the essay.

## 37d Latin abbreviations

| ABBREVIATION | LATIN | ENGLISH |
|---|---|---|
| c. | *circa* | about |
| cf. | *confer* | compare |
| e.g. | *exempli gratia* | for example |
| et al. | *et alia* | and others |
| etc. | *et cetera* | and so forth |
| i.e. | *id est* | that is |
| vs. [OR v.] | *versus* | versus |
| NB | *nota bene* | note well |

Reserve Latin abbreviations for use in footnotes and bibliographies. Use the appropriate English equivalent in your academic writing (except within parentheses).

# 38 Numbers

## 38a Numbers versus words

In most academic writing, spell out numbers of one or two words; use figures for all other numbers and amounts.

➤ Judson found ~~35~~ thirty-five old photographs at a garage sale.

➤ Our city has ~~one hundred forty-three~~ 143 restaurants.

When you give more than one number in a sentence or paragraph and at least one of those numbers is expressed by figures, all numbers should be expressed by figures, regardless of other rules.

➤ Of the 3,779 seniors, ~~ninety-six~~ 96 were from New York, ~~ninety-eight~~ 98 from Florida, and 428 from California.

In much technical and business writing, use figures for all numbers over ten. In the sciences and engineering, use figures for zero to nine as well, especially for exact measurements.

When two separate numbers appear together, spell out one and use figures for the other for clarity.

➤ In our literature class, we wrote six ~~five~~ 5-page papers.

## 38b Conventional uses of numerals

**DATES**

July 19, 1994 32 BCE [OR 32 BC] 60 CE [OR AD 60]

**TIME**

5:30 a.m. 3 p.m.

**ADDRESSES**

328 West 128th Street

**PAGE AND PART NUMBERS IN BOOKS**

page 67 chapter 2 volume 1

PLAYS
act 4, scene 2 [OR act IV, scene ii]

FRACTIONS, DECIMALS, AND PERCENTAGES
33 1/3  3.26 kilometers  75 percent

SCORES AND STATISTICS
The score was Cleveland 8, Minnesota 3. [OR 8 to 3, OR 8–3]

Surveys show that 7 runners out of 10 prefer Nike.

EXACT MEASUREMENTS OR COUNTS
6.32 miles  1.87 pounds  32,753 men

EXACT AMOUNTS OF MONEY
$19.99  $20  $56,007.68

LARGE NUMBERS
4.7 million  263,000,000  35 billion

IDENTIFICATION NUMBERS
Channel 3  Interstate 91  #10772683

## 38c Numbers at the beginnings of sentences

One hundred fifty-seven
➤ ~~157~~ passengers were stranded at the airport.

Or you can revise the sentence so that the number does not appear at the beginning.

➤ There were 157 passengers stranded at the airport.

# 39 Italics/Underlining

Italics, or slanted characters, are used, by convention, for the titles of certain works of art and, because of their visual contrast, to give emphasis to parts of written messages. In a handwritten or typed paper, continuous underlining indicates italics.

## 39a Titles of long works

BOOK *Blue Highways*

PLAY *The Glass Menagerie*

FILM *A Beautiful Mind*

RADIO OR TELEVISION SERIES *The West Wing*

COMIC STRIP *Dilbert*

MAGAZINE *Portable Computing*

JOURNAL *The American Scholar*

NEWSPAPER *The Village Voice*

LONG POEM *Paradise Lost*

LONG MUSICAL COMPOSITION *Phantom of the Opera*

PAINTING OR SCULPTURE *Mona Lisa*

COMPACT DISC OR CASSETTE TAPE *Abbey Road*

Enclose titles of shorter works such as poems of no more than a few pages; essays; and articles from newspapers, magazines, or scholarly journals in quotation marks. Do not underline (italicize) or put in quotation marks the Bible, books of the Bible, titles of legal documents, or the titles of your research papers and essays.

## 39b Names of ships, planes, trains, and spacecraft

*Delta Queen*
*Orient Express*
*Challenger*

*Spirit of St. Louis*
*Viking 7*
*Mir*

## 39c Numbers, letters, and words referred to as such or used as illustrations

- A large gold *10* was painted on the door.
- The letter *e* is the most frequently used letter in the English language.
- Did you know the word *piano* is a shortened form of *pianoforte*?

## 39d Foreign words and phrases

Whether you should italicize (underline) a foreign term depends on how widely speakers of English use it. Many foreign words have become part of the English language and do not, therefore, take italics (underlining). Consult your college dictionary if you are unsure.

➤ pax vobiscum [Latin: "peace be with you"]

➤ mano a mano [Italian: "hand to hand"]

➤ chérie [French: "dear one, darling"]

➤ café au lait [NOT café au lait]

➤ bon voyage [NOT bon voyage]

➤ du jour [NOT du jour]

## 39e For emphasis

For emphasis, writers sometimes underline (italicize) the key word in their argument.

➤ The time to face many of the problems of divorce is before marriage.

—Louise Montague

Underlining for emphasis can be effective but only when it is used sparingly.

# 40 The Hyphen

The hyphen is a versatile mark. Use it to form compound words; to add certain prefixes, suffixes, and letters to words; and to divide words. Do not confuse the hyphen (-) with the dash (-- or —). (See pages 109–110.)

## 40a Hyphen with compound words

Some compound words are two separate words (*half brother*); some are one word (*stepmother*); and some are two or more hyphenated words (*half-moon, father-in-law*). Sometimes the only way to be sure which form is correct is to consult your college dictionary. If the term does not appear as an entry in your dictionary, treat the compound as two words.

## 40b Hyphen with compound adjectives

When you put two or more words together to function as an adjective before a noun, hyphenate them; when you use these words after a noun, do not hyphenate them.

HYPHENATED FORMS

nineteenth-century literature

all-but-indestructible toy

NONHYPHENATED FORMS

the literature of the nineteenth century

The toy is all but indestructible.

Do not hyphenate when the first word of a descriptive phrase is an adverb ending in *-ly: widely known fact.*

Use suspended hyphens with a series of single-hyphenated modifiers before a noun.

➤ At Disney World, visitors buy one-, two-, or three-day passes.

## 40c Hyphen with compound numbers and fractions

Hyphenate compound numbers from twenty-one through ninety-nine and spelled-out fractions.

twenty-five     three-fourths

## 40d Hyphen with prefixes and suffixes

Hyphenate with the prefixes *all-*, *ex-* ("former"), *great-* (referring to a relative, e.g., *great-aunt*), and *self-*, and with the suffix *-elect.*

all-city     self-appointed     governor-elect

Also use a hyphen between a prefix and a proper noun or proper adjective.

pre-Columbian     un-American     anti-Semitic

And use a hyphen to join a single letter to a word.

U-turn L-shaped T-square

Hyphens prevent confusion between words otherwise spelled the same.

recount [to narrate the facts]

re-count [to count again]

Hyphens prevent difficulty in reading doubled or tripled letters that result from adding some prefixes and suffixes.

de-escalate anti-inflammatory ball-like

## 40e Hyphen to signal that a word is divided and continued on the next line

➤ Candidates who want respect should take firm positions on such controversial issues as campaign financing, unemployment, and the national debt.

In general, avoid hyphenating at the end of a line, but when there is not enough space for the full word without violating the right-hand margin, refer to the following seven rules for correct hyphenation:

1. Divide words between syllables.
2. If you need to break a hyphenated *compound* at the end of the line, break the word after the given hyphen. If you cannot follow this rule, carry the entire word to the next line.
3. If a consonant has been doubled in adding *-ing* to a word, divide between the double letters.
4. Do not divide one-syllable words.
5. Do not divide a word so that one letter is left at the end of a line or fewer than three letters carry over to the beginning of the next line; carry the entire word to the next line.
6. Do not divide proper names or the first name and middle initial.
7. Do not separate a contraction.

# 41 Spelling

Spelling problems differ from one writer to the next. Recognizing the nature of your own misspellings will help you develop corrective strategies.

## 41a Basic spelling rules

*ie/ei*

Use the traditional rhyme for *ie* and *ei* spellings.

| | |
|---|---|
| Write *i* before *e* | (believe, grieve, yield) |
| Except after *c* | (ceiling, deceit, receive) |
| Or when sounded like *ay* | (reign, veil, vein) |
| As in *neighbor* or *weigh*. | |

EXCEPTIONS either, financier, foreign, forfeit, leisure, neither, seize, sovereign, weird.

Final *e*

Drop a final *e* before adding a suffix beginning with a vowel but not before a suffix beginning with a consonant.

| | |
|---|---|
| fame + -ous = famous | achieve + -ment = achievement |
| desire + -able = desirable | love + -ly = lovely |
| hope + -ing = hoping | hope + -ful = hopeful |

EXCEPTIONS acknowledgment, argument, dyeing, hoeing, judgment, truly

Retain the final *e* after a soft *c* or *g*.

courage + -ous = courageous

exchange + -able = exchangeable

notice + -able = noticeable

Final consonant before suffix beginning with vowel

Double a final consonant before a suffix beginning with a vowel if the word is one syllable and ends in a consonant preceded by a single vowel.

swim + -ing = swimming

fit + -ed = fitted

set + -ing = setting

Double a final consonant before a suffix beginning with a vowel if the word ends in an accented syllable and a consonant preceded by a single vowel.

commit + -ing = committing

occur + -ed = occurred

abhor + -ent = abhorrent

**EXCEPTIONS** feeding, benefited

### Final *y*

Change a final *y* preceded by a consonant to *i* before adding a suffix other than *-ing*. Keep the final *y* when it is preceded by a vowel.

beauty + -ful = beautiful
embody + -ment = embodiment
modify + -er = modifier
lonely + -ness = loneliness

array + -ed = arrayed
buy + -er = buyer
obey + -ing =obeying
sway + -ed = swayed

**EXCEPTIONS** daily, laid, paid, said

### Noun plurals

Most nouns form the plural by adding *-s* or *-es* to the singular, but other rules apply as well.

| RULE | EXAMPLES |
|---|---|
| Add *-s* to most nouns | stick + *-s* = sticks<br>pattern + *-s* = patterns<br>elephant + *-s* = elephants |
| Add *-es* if the noun ends in *s*, *sh*, *ch*, *x*, or *z*. | pass + *-es* = passes<br>brush + *-es* = brushes<br>peach + *-es* = peaches<br>tax + *-es* = taxes<br>buzz + *-es* = buzzes |

(*continued*)

(*continued*)

| RULE | EXAMPLES |
|---|---|
| For nouns ending in *y* preceded by a consonant, change the *y* to *i* and add *-es*. | spy + *-es* = spies<br>inventory + *-es* = inventories<br>penny + *-es* = pennies |
| For most nouns ending in *o* preceded by a consonant, add *-es*. | tomato + *-es* = tomatoes<br>echo + *-es* = echoes<br>embargo + *-es* = embargoes<br>EXCEPTIONS autos, memos, pros |
| Some nouns ending in *f* or *fe* form the plural by adding *-s*, whereas others change the *f* to *v* and add *-es*. | roof + *-s* = roofs<br>chief + *-s* = chiefs<br>safe + *-s* = safes<br>wharf + *-es* = wharves<br>wife + *-es* = wives<br>hoof + *-es* = hooves |

Some nouns form their plurals by changing internal vowels.

| SINGULAR | PLURAL |
|---|---|
| foot | feet |
| mouse | mice |
| woman | women |

Plural forms of foreign borrowings usually retain the plural of the original language.

| SINGULAR | PLURAL |
|---|---|
| datum | data |
| tableau | tableaux |
| appendix | appendices [OR appendixes] |

Some foreign borrowings, however, have an Anglicized plural in addition to the plural of the original language; both forms are acceptable.

| SINGULAR | PLURAL | |
|---|---|---|
| formula | formulae | formulas |
| index | indices | indexes |
| syllabus | syllabi | syllabuses |

Some nouns make no changes to form the plural.

| SINGULAR | PLURAL |
|---|---|
| deer | deer |
| fish | fish [OR fishes] |
| swine | swine |

Form the plurals of compound words by adding *-s* to the end of the word except when the first word is more important.

| SINGULAR | PLURAL |
|---|---|
| mastermind | masterminds |
| passer-by | passers-by |
| editor in chief | editors in chief |

## 41b Words that sound alike but have different meanings and spellings

Words—such as *cite, site,* and *sight,* or *stationary* and *stationery*—that sound alike but have different spellings and meanings are called *homophones.* Some frequently confused words are listed here:

| | |
|---|---|
| affect | *v.*, to influence |
| effect | *n.*, result |
| ascent | *n.*, movement up |
| assent | *n.*, agreement; *v.*, to agree |
| complement | *n.*, that which completes |
| compliment | *n.*, praise |
| council | *n.*, assembly |
| counsel | *n.*, lawyer; *v.*, to advise |
| eminent | *adj.*, distinguished |
| immanent | *adj.*, remaining within |
| imminent | *adj.*, about to happen |
| past | *n.*, time gone by |
| passed | *v.*, past tense of *pass,* to go by |
| principal | *n.*, school official; in finance, a capital sum; *adj.*, most important |
| principle | *n.*, basic law or rule of conduct |
| steal | *v.*, to take without permission |
| steel | *n.*, metal; *adj.*, referring to metal |

| | |
|---|---|
| than | *conj.*, used in comparisons |
| then | *adv.*, at that time |
| vain | *adj.*, futile; excessively proud |
| vane | *n.*, object that indicates wind direction |
| vein | *n.*, blood vessel |

## 41c Commonly misspelled words

Following is a list of words that college students frequently use and commonly misspell.

COMMONLY MISSPELLED WORDS

absence
academic
accidental
accommodate
acknowledge
acquaintance
acquire
across
address
all right
a lot
altogether
amateur
analyze
answer
apparently
appearance
appropriate
argument
arrangement
ascend
athletic
attendance
audience

basis
beginning
belief
beneficial
bulletin
bureau
business

calendar
candidate
changeable
characteristic
column
committee
competitive
concede
conceivable
conferred
conscience
conscientious
conscious
courteous
criticism
curiosity

decision
definitely
describe
description
desperate
disappear
disastrous
dissatisfied

eighth
eligible
embarrassment
eminent
entirely
environment
especially
exaggerate
exhaust
existence

familiar
fascinate
foreign
forty

government
grammar
guidance

harass
height
humorous
hypocrisy

illiterate
incidentally
incredible
indispensable
inevitable
intelligence
interesting

irrelevant
irresistible

jealousy

knowledge

laboratory
license

maneuver
mathematics
mischievous

necessary
noticeable

occasion
occurrence
omitted
optimistic

pamphlets
parallel
particular
pastime
perseverance
physical
playwright
politics
practically
precede
precedence
preference
prejudice
privilege
proceed
pronunciation

quiet
quite

recommend
reference
referred
repetition
restaurant
rhythm
ridiculous

sandwich
schedule
secretary
seize
separate
sergeant
similar
sincerely
sophomore
succeed
supersede
surprise
subtly

thorough
tragedy
truly

unnecessarily
usually

vacuum
vengeance
villain

weird
whether
writing
written

# *Design*

# 42 Basic Page Design

Design decisions are closely tied to the subject and purpose of your writing. What is the subject matter? A research paper in the humanities, for example, must follow MLA guidelines (see pages 196–202). How long is the paper? Would internal headings make it more readable? What is your purpose, and who is your audience? Is your topic very complicated? Would a list instead of an extended paragraph help your reader understand a particular point better? Aside from these questions, there are, of course, such fundamental matters as setting margins and line spacing and choosing a text typeface and type size. With all the capabilities available on even the most basic computer software, you have many options from which to choose.

## 42a Page design

For academic purposes, use 8 1/2" × 11", twenty-pound white paper, and a letter-quality printer. Formatting of any writing begins with four basic decisions: margins, line spacing, type style, and type size.

### Margins and line spacing

Unless your instructor tells you otherwise, leave a one-inch margin at the top and bottom of the page and on the right and left sides. Leave a ragged (uneven) margin on the right, because a justified (even) margin causes odd spacing between words. Turn off the automatic hyphenation function on your computer to avoid hyphenated words at the ends of lines.

Most academic writing is double-spaced. Memos and business letters are usually single-spaced to ensure they fit on a single page, the preferred length in the business world.

### Type styles and sizes

Use familiar styles (Courier, Times Roman, Helvetica, Geneva) in a standard ten- or twelve-point size. Do not use a script typeface or all italics or capitals for your main text; they are too difficult to read. For headings and subheadings, use the same type style, though you may slightly increase the point size.

When you have made your basic page design choices, you may want to print out a sample page. Adequate margins and a type style and size that do not call attention to themselves will transmit your message most effectively.

## 42b Improving readability

### Headings and subheadings

There is no substitute for good organization in a piece of writing, but in the case of long or complex research papers and reports, headings and sometimes even subheadings can enhance readability. Headings and subheadings break an extended piece of writing into visibly distinct sections, allowing the reader to enter and exit your train of thought and to stand back and see what has gone before and what is to come.

Consistency is very important in using headings. First, you must be consistent in how you phrase headings. Most headings are a single word, usually a noun (e.g., Gardens) or phrase (e.g., Types of Gardens). Often they are gerund phrases (e.g., Mulching the Garden). Headings may also be questions, which are then answered in that section (e.g., When Should You Prune Roses?). Whichever style heading you start with, however, you must continue using it throughout your paper. If you are using both headings and subheadings, you can make all your headings one style (say, single-word nouns) and all your subheadings another style (perhaps gerund phrases).

You must also be consistent in the typeface and size of your headings. Suppose your text type is ten-point Times Roman. All your headings should also be Times Roman, but you might put your main headings in fourteen point and your subheadings in twelve point. You have other options in how you present your headings: boldface (heavier type), underlining, italics or bold italics, all capitals, and capitals and lower case.

**Weeding Gardens** [bold]
<u>Weeding Gardens</u> [underlined]
*Weeding Gardens* [italics]
***Weeding Gardens*** [bold italics]
WEEDING GARDENS [all caps]

As with phrasing, you may choose one style for headings and one for subheadings. Again, though, you must be consistent in whichever you choose.

Finally, you must be consistent in the placement of headings. By convention, main headings are usually centered, and subheadings are at the left margin.

Main Heading

Subheading

Headings can improve the readability of a long or complex piece of writing, but they must be used wisely; too many headings and levels of subheadings can be distracting. Every heading should introduce a major concept, not a minor point.

### Lists

You can also help your reader by pulling material out of the text and presenting it as a more visually accessible list. Some material lends itself to list making more than others, of course. Steps in a process, categories, and how-to instructions naturally fit into a list.

Set off a list by indenting from the left margin. You can draw attention to the individual items in your list with graphic symbols, such as bullets (large, solid dots •), squares (■), or dashes (—). Graphic symbols are a good idea if the items in your list run over one line. See how graphics work in this example:

Experts say there are four major ways to relieve stress:

- consider whether you are accurately appraising the situation;
- use your problem-solving skills to take action;
- seek the support of family and friends; and
- pay attention to your health by eating right, getting enough exercise, and going to bed at a regular time.

This example also shows that you should introduce a list with a colon, put a semicolon or comma after each item, and end with a period. Note, too, that the phrasing should be parallel. In this list, every item begins with a verb: *consider, use, seek, pay.*

## 42c Using visuals

Visuals—charts and graphs, tables, diagrams, illustrations and photographs, and clip art—can add a new dimension to your writing. They allow a reader to visualize as well as read the ideas and information you are presenting.

Charts and graphs show relationships among numerical data. A pie chart is a circular graph; it shows a whole and the percentages that make up the whole. A bar graph uses bars, usually vertical, to indicate frequency or quantity; it is useful for showing large-scale size or mass comparisons. A line graph shows finely delineated data—such as age, speed, or temperature—in a continuum. (See page 172.) Overlaid lines in a line graph highlight points of intersection and deviation. The same data—total units sold—is used in the following examples to show that the information you want to present determines the graph you must use. (See also page 172.)

A table is useful for summarizing large amounts of information, either numerical data or text narrative. Because tables are columnar, they are also useful for making comparisons. (See the table of Latin abbreviations on page 121.)

A diagram, or drawing, is a concise visual representation of an object or idea that would take many words to convey. The diagram of business letter formats on page 144, for example, allows the reader to see quickly the three different formats available.

PIE CHART

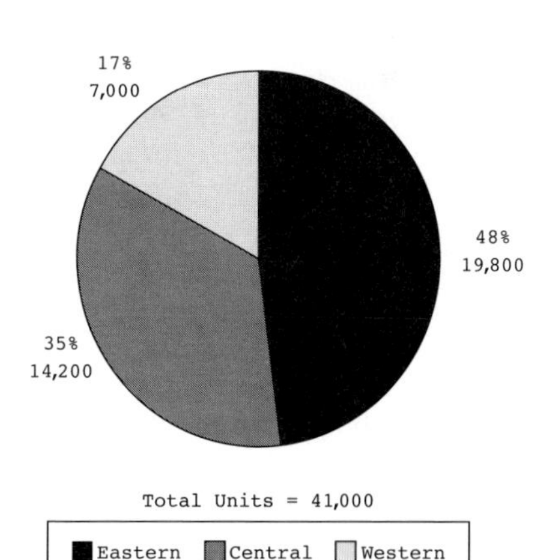

BAR GRAPH

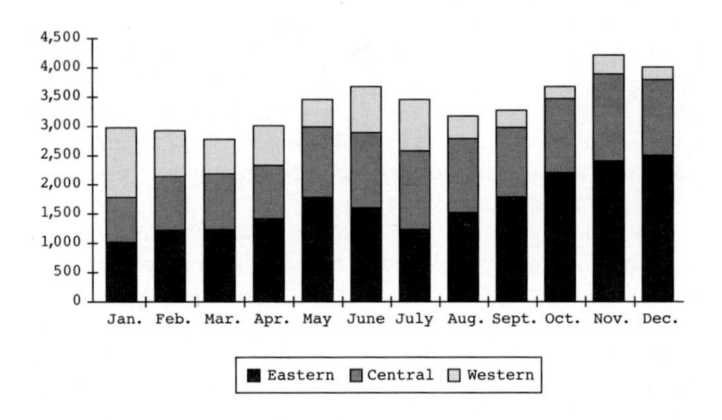


Photographs, especially in posters, brochures, and newsletters, can be very effective in capturing readers' attention, adding information, and enlivening a document. They are not used frequently in academic writing, but you can enhance your documents with images of people, places, and things. For example, the photograph on page 141 of a Florida election official puzzling over a disputed ballot in the 2000 presidential election communicates far more than words alone can do.

Clip art is available on most word-processing programs, and much more of every conceivable type and description exists on the Web. Often it is available for no charge. Simply use your search engine to look for the images you need, and copy and paste them into your documents where appropriate.

You can create your own visuals with computer graphic software, or you can use visuals you find in your research. If you use visuals someone else created, you must give proper credit. The acknowledgment should be placed directly beneath the visual and should include the name of the creator and the place and date of publication. The acknowledgment should also be included in your list of works cited. Be sure that your visuals are properly sized (neither too large to fit within your margins nor too small to be legible), that they

PHOTOGRAPH

(Braley/Reuters/TimePix. Contributor: TimePix. Reprinted by permission.)

are placed close after the text to which they correspond, and that the quality of your reproduction does not defeat the purpose of the visual aids. It's better not to use a visual at all than to use one that confuses rather than enhances your message.

CLIP ART

# 43 Business Correspondence

## 43a Business letters

In the past, business letters were formal and stiff, with standardized phrasings and exaggerated courtesy. Today, courtesy is just as important in a business letter, but modern business practice allows less formality. Clarity is imperative; there should be no confusion over the message you are sending, and your writing should be grammatically correct and free of clichés, jargon, and other inappropriate language. Finally, business letters follow conventional formats.

### Format

Three formats are used in business letters today: full block, modified block, and indented. Select the format appropriate for your purpose: full block is most formal, whereas indented is the least formal. A job application letter, for example, is best presented in the full-block format.

All business letters should be typed on unlined white paper and on one side of the page only. Use one-inch margins on the left and right sides. Single-space the letter except where indicated in the accompanying sample layouts of the three formats. A sample letter of application in the full-block pattern is shown on page 145.

### Return address

Unless you are using letterhead stationery on which your name and address or your company's name and address are imprinted, include your return address. Always include the date. Placement depends on the format you have chosen (see the sample layouts on page 144).

### Inside address

In all formats, type at the left margin the person's name, title, name of department if any, name of the business or organization, and address of the person to whom you are writing. If you are addressing a woman and do not know the way she prefers to be addressed, use *Ms.* If a title is necessary and it is short, put it after the person's name, separated by a comma. Long titles should be placed on the next line. Capitalize the title in both cases.

### Salutation

If you do not know the name of the person you are addressing, use *Dear Sir* or *Madam,* the person's position (*Dear Manager* or *Dear Director of Personnel*), the name of the company (*Dear Klein Refrigeration*), or the subject of your letter (*Recommendation for James M. Morton*). A colon follows the salutation.

### Body

The body of a business letter begins at the left margin in the full-block and modified-block formats. If you are using the indented format, indent each paragraph five spaces.

### Close

The first word (and *only* the first word) of the close begins with a capital letter. Customary closes include *Sincerely, Sincerely yours,* and *Yours truly.* Put a comma after the close. The placement of the close depends on the format (see the sample layouts on page 144).

### Name and signature

Four lines beneath the close, type your name (and your title, if you have one, on a separate line), and then place your signature above the typed name (see the sample letter of application on page 145).

## SAMPLE BUSINESS LETTER FORMATS

Full block

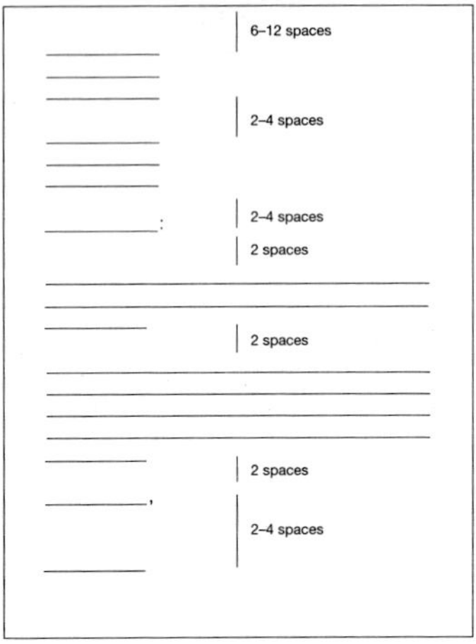

6–12 spaces
2–4 spaces
2–4 spaces
2 spaces
2 spaces
2 spaces
2–4 spaces

Indented

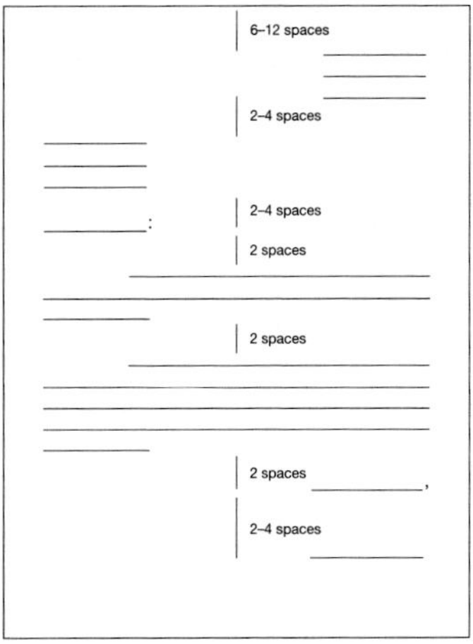

6–12 spaces
2–4 spaces
2–4 spaces
2 spaces
2 spaces
2 spaces
2–4 spaces

Modified block

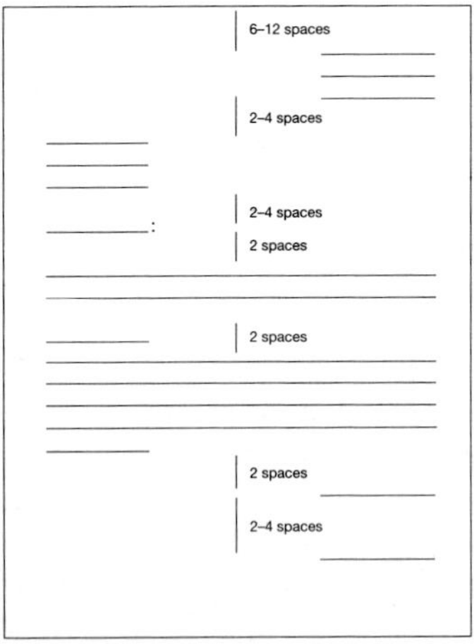

6–12 spaces
2–4 spaces
2–4 spaces
2 spaces
2 spaces
2 spaces
2–4 spaces

## SAMPLE LETTER OF APPLICATION: FULL-BLOCK PATTERN

512 Living/Learning Center
University of Vermont
Burlington, VT 05401
November 1, 2002 — **Return address and date**

Mr. Harold Sampson, Director
Management Training Program
Acme Insurance Group
100 Asylum Street
Hartford, CT 06100 — **Inside address**

Dear Mr. Sampson: — **Salutation**

**Body**

Through the Career Development Office of the University of Vermont, I have learned that you have openings in your Management Training Program. Your program interests me very much, and I would like to apply for a position in it.

On May 12, 2003, I will graduate from the School of Business Administration with a major in insurance. I believe that the number and variety of courses I have taken in both insurance and marketing qualify me to apply for your program.

In addition to my coursework, I have worked summers and part-time during the school year as a clerk for Mr. Hugh Johnson of your Burlington office. I have also accompanied Mr. Johnson at various times when he called on clients. I believe, therefore, that I have practical experience as well as theoretical experience on which to base my choice of career in the insurance management field.

My résumé is enclosed. I can also provide you with the names of individuals who can give you assessments of my efficiency and dependability as a worker.

I would be pleased to travel to Hartford to meet with you should you think it necessary. You may call me at (802) 555-1234 or write me at the above address or at <jmorton@zoo.uvm.edu>.

Sincerely yours, — **Close**

*James M. Morton* — **Signature**

James M. Morton

encl. résumé
cc: Career Development Office — **Notations**

## 43b Résumés

A résumé tells a prospective employer of your qualifications for a particular position. Because your chances of getting a job interview—the first step in getting a job—rest, in large part, on how well your résumé communicates your training and experience, your skills and abilities, you will want to think carefully about what information to include and how to present it. Your résumé should reveal why you are the ideal candidate for the position.

In preparing your résumé and getting ready for a job search, you may find it helpful to read *What Color Is Your Parachute?* by Richard N. Bolles (Ten Speed Press), an excellent job search guide with additional online information available at <http://www.jobhuntersbible.com>. Also, consult with your college's placement office.

A résumé follows a conventional format and includes the following categories of information. (See a sample résumé on page 148.)

### Name, addresses, phone numbers

Center your name at the top of the résumé. Give both your present address and permanent home address, an e-mail address, and voice and fax phone numbers, including area codes.

### Job objective, position sought

As concisely as possible, describe the position you are seeking.

### Education

Give the name of the institution you graduated from, the year you received your degree, specific degree, your major and minor, and, if you like, your grade point average. Include other educational experiences if they are pertinent to the position you are seeking.

### Awards and honors

Include awards and honors that reveal you are a superior student or have special talents or leadership abilities.

Employment

List the jobs you have held, starting with the most recent. If your responsibilities are not clear from your job titles, briefly outline what they were. Include the name of your supervisor.

Activities

List the campus organizations, volunteer groups, and collegiate sports in which you participated, and name or describe the positions of responsibility you held.

Skills

List the computer operating systems and programs with which you are familiar. If you have desktop-publishing experience or have worked with graphics or cameras, add that information. Other skills might include sales, telemarketing, academic research, laboratory research, political campaigning, data collection and entry, programming, and science or social science fieldwork.

References

You can either indicate where your references are on file (if your institution provides this service) or list the names, addresses, and phone numbers of your references. Before listing a person's name as a reference, be sure to get permission.

## SAMPLE RÉSUMÉ

**James Michael Morton**

Present address:
512 Living/Learning Center
University of Vermont
Burlington, VT 05401
(802) 555-1234
Fax: (802) 555-3055
jmorton@uvm.edu

Permanent address:
1065 Peer Avenue
Clifton, NJ 07013
(201) 555-1212

**Job Objective** Position in a management training program in the insurance industry with the goal of becoming an insurance executive

**Education**

1999–2003 B.S., May 2003, University of Vermont (expected)
Major: insurance
Minor: marketing
Grade point average: 3.50

**Awards, Honors** Dean's list for last four semesters
Recipient of the Dean's Scholarship, School of Business Administration, senior year

**Employment**

2001–2002 Part-time employment during the academic year: office worker and sales assistant for Acme Insurance Group, Burlington, Vermont. Supervisor: Mr. Hugh Johnson

1999 Summer: Office assistant for Acme Insurance Group, Burlington, Vermont
Supervisor: Mr. Hugh Johnson

1998–2000 Summer: Proofreader for the Burlington Free Press

**Activities** Sigma Phi Epsilon, Business Manager (2000), President (2002–2003); intramural volleyball; peer tutor

**Skills** MS-DOS, WordPerfect, Adobe Pagemaker, Microsoft Office, Lotus 1-2-3

**References** Center for Career Development
University of Vermont
Burlington, VT 05405
(802) 555-3450

## 43c Memos

A memo is a brief, usually one-page, written communication between workers within the same office or company. Memos are used for a wide variety of communications: queries and answers; reminders; calls for meetings; and announcements affecting products, services, and personnel, for example. It is a good idea to cover only one topic in a memo so that the recipient can take action on it if necessary and then put it in the proper file.

Memos follow a conventional format. A company or organization may even supply special printed stationery or word-processing templates with the company logo at the top and the headings

To:

From:

Date:

Subject:

You can add a *cc: (copy)* to send a copy to someone after the *To:* or at the very end of the memo.

In writing a memo, use your own voice. Be natural but to the point. Courtesy is also important; use *please* and *thanks.* Sign your name or initials next to your name or at the end of the memo. A sample memo is shown on page 150.

SAMPLE MEMO

MEMO

University of Vermont
Department of English

To: Members of the Department of English
From: Alan Broughton AB
Date: April 27, 2002
Subject: Move to University Heights

Here is more information concerning our move.

Before the move, we will have at least one work-study student to help us. This student's primary duty will be to help pack the main office, but the student will also be available to help with your needs.

Here is the moving schedule:
- week of May 3—classroom furniture (so we will be ready for the opening of summer school on May 10);
- week of May 10—main office;
- week of May 17—faculty offices, on a floor-by-floor basis.

Computers will be moved by the Computer Technology office. This will require careful coordination with the moving company so that only one group is working in an office at a time.

Movers will transport furniture, files, and books. Plan to take personal or fragile objects home and move them into your new office yourself.

I'll keep you informed as I learn more. Meanwhile, thanks for your continued good humor during this hectic time.

# *Researched Writing*

## 44 Using Print Sources

In most cases, you should use print sources (books, newspapers, magazines, scholarly journals, encyclopedias, pamphlets, brochures, and government publications) as your primary tools for research. Print sources, unlike Internet sources, are either reviewed and refereed by experts in the field, approved and overseen by a reputable publishing company or organization, or examined by editors and fact checkers for accuracy and reliability. Unless you are instructed otherwise, you should always use print sources in your research.

To find print sources, search through your library's reference works, card or computer catalog, periodical indexes—both print and electronic—and other electronic databases. They will enable you to generate a working bibliography, a preliminary listing of books, magazine and newspaper articles, public documents and reports, and other sources that sound as if they might help you explore or answer your research question or objective. At this early stage, it is better to err on the side of listing too many sources rather than finding yourself at a later stage trying to relocate those you discarded too hastily. If you have questions while working in the library, ask the reference librarian for assistance.

## 45 Using Internet Sources

Internet sources can be informative and valuable additions to your research; for example, you might find a just-published article from a university laboratory or a news story in your local newspaper's online archives. Generally, however, you should use Internet sources alongside print sources and not as a replacement for them. Whereas print sources are published under the guidance of a publisher or organization, practically anyone with access to a computer and a modem can put text and pictures on the Internet; there is no governing body that checks for content or accuracy. The Internet offers a vast number of useful, legitimate, and carefully maintained resources, but it also contains many bogus facts and many examples of rumor, conjecture, and unreliable information. It is your responsibility to evaluate whether a given Internet source should be trusted.

Use search engines to find Internet sources. You should try more than one search engine in your research because each one yields slightly different results. You can access search engines easily by using the Net Search function on your browser.

Following is a list of popular search engines:

- *37.com* **<http://37.com>**

  Searches up to thirty-seven other search engines. Also allows you to select a particular engine. Fast and powerful.
- *All-in-one* **<http://www.albany.net/allinone>**

  Searches all other search engines for Web sites. A fill-in search form is provided for your requests.
- *Alpha Search* **<http://www.calvin.edu/library.as>**

  A subject catalog of Internet gateways, which are also known as "webliographies," "meta-indexes," and in the paper world "subject bibliographies." All sites are large, academic in nature, and active.
- *Alta Vista* **<http://www.altavista.digital.com>**

  Fast and comprehensive, this search engine indexes millions of sites and is, therefore, useful for retrieving well-narrowed or hard-to-find topics.
- *Dogpile* **<http://www.dogpile.com>**

  Allows a search using multiple search engines at the same time.
- *Excite* **<http://www.excite.com>**

  Searches over 250 million indexed Web pages and multimedia items.
- *Google* **<http://www.google.com>**

  Unlike other search engines. *Google* searches with keywords using a powerful ranking search logic.
- *HotBot* **<http://www.hotbot.com>**

  Uses both form-based and limited expert query modes. Allows users to save searches and personal settings.
- *InfoSeek* **<http://guide.infoseek.com>**

  Uses keyword, concept, and related terms. Recommended for Usenet newsgroups searches.

- *Lycos* **<http://www.lycos.com>**

  Indexes over 90 percent of Web sites. Lycos allows you to search FTP and gopher sites but not Usenet newsgroups.
- *MetaCrawler* **<http://www.metacrawler.com>**

  *MetaCrawler* does not maintain a local database but rather relies on the databases of various Web-based sources. It sends your queries to several Web search engines, including *Lycos, Infoseek, WebCrawler, Excite, AltaVista,* and *Yahoo!*. It then puts the results into a uniform format and ranks them by relevance.
- *Netscape's Net Search* **<http://home.netscape.com/home/internetsearch.html>**

  Easily accessed on *Netscape*'s taskbar, *Net Search* is a jump station to Web search engines and useful in beginning a search.
- *Northern Light* **<http://www.northernlight.com>**

  Builds customized folders by topic and Web address. It also provides citations to articles from thousands of journals and newspapers.
- *WebCrawler* **<http://wc3.webcrawler.com>**

  One of the fastest search engines because it is selective. Search results are ranked by frequency of occurrence of keyword.

In addition to the subject directories provided on the home pages of most search engines, a number of Web sites contain more comprehensive and sophisticated subject directories. The following "stand-alone" directories can be helpful to your research. They can be reached at their Web addresses.

- *The Argus Clearinghouse* **<http://www.clearinghouse.net>**

  A directory of detailed subject guides, each containing links compiled by experts in their fields. An excellent place to begin your research.
- *Galaxy* **<http://www.einet.net>**

  An extensive database of subject-classified links with descriptive titles in lieu of annotations.
- *INFOMINE: Scholarly Internet Resource Collections* **<http://lib-www.ucr.edu>**

  Nearly ten thousand subject-classified links to Internet sites, helpful to both students and scholars.

- *The Internet Public Library Reference Center* **<http://www.ipl.org/ref>**

  Librarians have annotated this list of subject-classified links provided by the School of Information at the University of Michigan.

- *Liszt* **<http://www.liszt.com>**

  A well-managed database of discussion lists that also provides useful information on how to select and subscribe. Searchable by category and topic.

- *Scholarly Journals Distributed by the World Wide Web,* by Robert C. Spragg **<http://info.lib.uh.edu/wj/webjour.html>**

  An alphabetical listing of links to scholarly journals on the Internet.

- *U.S. Federal Government Agencies: A List of Federal Agencies on the Internet* **<http://www.lib.lsu.edu/gov/fedgov.html>**

  A hierarchically arranged list of Internet links to government agencies and their subagencies.

- *The World Wide Web Virtual Library Subject Catalog* **<http://www.w3.org/pub/DataSources/bySubject/overview.html>**

  A large directory of guides to subject-classified links with limited annotations.

- *Yahoo!* **<http://www.yahoo.com>**

  A huge, regularly updated database of subject-classified sites with informative annotations. Searchable by keywords in URLs, titles, or descriptions.

# 46

# Using Directory and Keyword Searches

## 46a Using subject directories to refine your research topic

The subject directories on the home pages of search engines make it easy to browse various subjects and topics, a big help if you are undecided about your exact research question or simply want to see if there is enough material to supplement your research work with print sources. Often the most efficient approach to Web research is to start with the subject directory provided by most search engines. Once you choose a subject in the directory, you can narrow the subdirectories and eventually arrive at a list of sites closely related to your topic.

Suppose you are researching integration in major league baseball in the 1940s, and you are using the search engine *Google*. This is the subject directory that would appear on your screen:

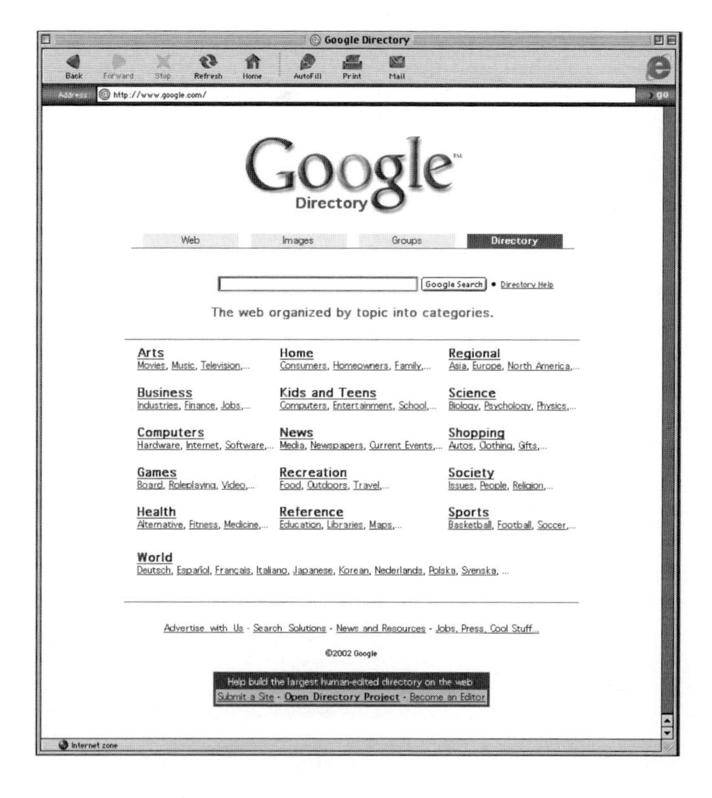


You would click on "Sports." This link takes you to a screen that lists a hundred individual sports. For "Baseball" alone there are 6,223 Web sites.

Adventure Racing (104)
Airsoft (81)
Animal Fighting (14)
Archery (230)
Badminton (107)
**Baseball** (6223)
Basketball (9863)
Billiards (220)
Bocce (25)
Goalball (11)
Golf (4025)
Greyhound Racing (74)
Gymnastics (1218)
Handball (16)
History (33)
Hockey (7265)
Horse Racing (925)
Humor (62)
Rope Skipping (21)
Rounders (7)
Running (2871)
Sailing (1238)
Sepak Takraw (16)
Shooting (97)
Shopping (10616)
Skateboarding (744)
Skating (1138)

When you click on "Baseball," you arrive at a screen listing categories of sites related to the national pastime.

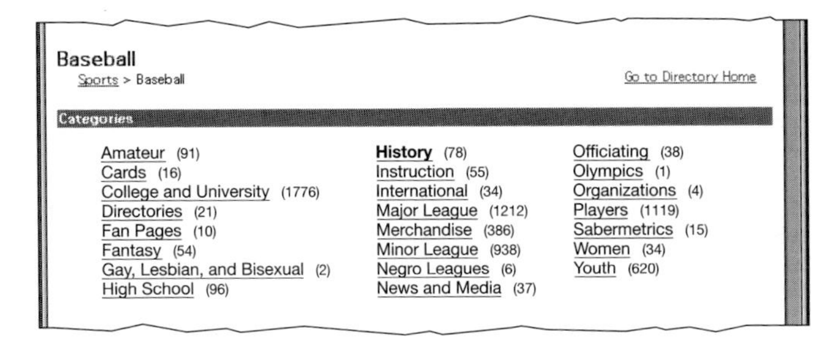


Because you are interested in events that occurred some time ago, the "History" option is a natural. Clicking on "History" takes you to a screen with a list of sites for baseball history, including "By Popular Demand: Jackie Robinson and Other Baseball Highlights."

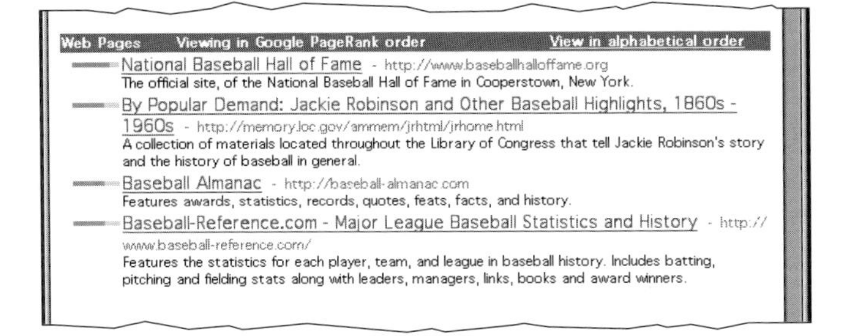


Clicking on this site takes you to the story of Jackie Robinson, the first African American to play major league baseball. Here you find links to "Baseball, the Color Line, and Jackie Robinson," "color line," "Negro Leagues," and "civil rights activities."

## 46b Using keyword searches to seek specific information

When you type in a keyword in the "Search" box on a search engine's home page, the search engine goes looking for Web sites that match your term. One problem with keyword searches is that they can often produce tens of thousands of matches, making it difficult to locate those of immediate value. For that reason, you need to make your keywords as specific as you can and make sure that you have the correct spelling. It is always a good idea to look for search tips on the help screens or advanced search instructions for the search engine you are using before initiating a keyword search. Once you start a search, you may want to narrow or broaden it, depending on the number of hits you get.

When using a keyword search, you need to be careful about what keywords will yield the best results. If your words are too general, your results can be at best unwieldy and at worst not usable at all. During an initial search for a paper on recycling old newspapers, one student typed in *recycling + newspapers*. To her surprise, this search produced more than 3.6 million hits. After some thought, and after learning about *Alta Vista*'s advanced search feature, she tried *recycling* + "*old newsprint*"—a search that yielded a far more manageable 489 hits. Among these hits, she located several Web sites that related directly to her project.

### Refining Keyword Searches

While some variation in command terms and characters exists among electronic databases and popular search engines on the Internet, the following functions are almost universally accepted. If you have a particular question about refining your keyword search, seek assistance by clicking on "Help" or "Advanced Search."

- Use quotation marks or parentheses to indicate that you are searching for the words in exact sequence: "*whooping cough*," *(Supreme Court)*.

- Use AND or a plus sign (+) between words to narrow your search by specifying that all words need to appear in a document: *tobacco AND cancer, Shakespeare + sonnet.*
- Use NOT or a minus sign (–) between words to narrow your search by eliminating unwanted words: *monopoly NOT game, cowboys – Dallas.*
- Use OR to broaden a search by requiring that only one of the words need appear: *buffalo OR bison.*
- Use an asterisk (*) to indicate that you will accept variations of a word: "*food label**" for food labels, food labeling, and so forth.

# 47 Selecting and Evaluating Sources

Your library work will involve working with print sources, electronic and CD-ROM indexes and databases, and online sources. In all cases, though, the process is essentially the same: you are winnowing from many sources the best ten to twenty-five, depending, of course, on the size of the project.

## 47a Preview your print and online sources

You will not have to spend much time in the library to realize that you do not have the time to read every print source that appears relevant. And the abundance of print sources housed in your library does not begin to approach the number of online sources available to you. Your library is local, but the Internet is a global network of computers and computer users. You can access information stored on computers anywhere in the world and communicate with people worldwide who might have information on your research topic.

Given the abundance of print and Internet sources, the key to successful research is identifying those books, articles, Web sites, and other online sources that will help you most. You must preview your potential sources to determine which you will read, which you will skim, and which you will simply eliminate. Use the following previewing strategies and suggestions to help you identify those sources that hold the most promise:

PREVIEWING A BOOK

Read the dust jacket or cover copy for insights into the book's coverage and the author's expertise.
Scan the table of contents, identifying chapters that address your research question.
Read the author's preface, looking for his or her thesis and purpose for writing the book.
Checking the index, locate key words or key phrases related to your research question.
Read the opening and concluding paragraphs of any promising chapters; if you are unsure, skim the whole chapter.
Does the author have a discernible bias?

PREVIEWING AN ARTICLE

What do you know about the journal or magazine publishing the article? What is its reputation, and who are its readers?
Consider the title or headline of the article as well as the opening paragraph or two and the conclusion.
For articles in journals, read the abstract if there is one.
Examine any photographs, charts, graphs, or other illustrations that accompany the article.

PREVIEWING A WEB SITE

Examine the home page of the site. Do the contents and links appear related to your research question?
Can you identify the author of the site? Are the author's credentials available?
Has the site been updated within the last six months? Look for this information at the bottom of the home page.

## 47b Evaluate your print and online sources

Before you begin to take notes, evaluate your sources for their relevance, bias, overall argument, and reliability in helping you answer your research question. Look for the writers' main ideas and arguments. Read critically, and, while it is easy to become absorbed in sources that support your own beliefs, always seek out sources with opposing viewpoints, if only to test your own position. And remember not to lapse into reading material not related to your topic just because it seems interesting. Look for information about the authors themselves, information that will indicate their authority and perspective or bias on the subject. And it is always helpful to know the reputation and special interests of book publishers and magazines, because you are likely to get very different views—liberal, conservative, feminist, international—on the same topic depending on the publisher or magazine you read. The Checklist for Evaluating Print and Electronic Sources will assist you.

### Checklist for Evaluating Print and Electronic Sources

1. What is the writer's thesis or claim?
2. How does the writer support this thesis? Does the evidence seem reasonable and ample, or is it mainly anecdotal?
3. Does the writer consider opposing viewpoints?
4. Does the writer have any obvious political or religious biases? Is the writer associated with any special-interest groups, such as Planned Parenthood, the Sierra Club, or the National Rifle Association?
5. Is the writer an expert on the subject? Do other writers mention this author in their work?
6. Does the publisher or publication have a reputation for accuracy and objectivity?
7. What is the author's purpose: to inform, or to argue for a particular position?
8. Do the writer's thesis and purpose clearly relate to your research question?
9. Does the source appear to be too general or too technical for your needs?
10. Does the source reflect current thinking and research in the field?

The quality of sources on the Internet varies tremendously. Because literally anyone can post anything on a Web site or to a newsgroup, it becomes your responsibility to assess the relevance, bias, and accuracy of each and every Internet source. Consider the following additional questions when evaluating Web sites.

## Questions for Evaluating Web Sites

**TYPE OF WEB PAGE**

Who hosts the Web site? Often the top-level domain name indicates the source of information provided. These are the most common top-level domain names:

| | |
|---|---|
| com | Business/commercial |
| edu | Educational institution |
| gov | Government |
| mil | Military |
| net | Various types of networks |
| org | Nonprofit organization |

**AUTHORITY/AUTHOR**

Is it clear what individual or company is responsible for the site?

Can you verify whether the site is sanctioned officially by an organization or company?

What are the author's or company's qualifications for writing on this subject?

Is there a way to verify the legitimacy of this individual or company (e.g., are there links to a home page or résumé)?

**PURPOSE AND AUDIENCE**

What appears to be the author's or sponsor's purpose in publishing this Web site?

Who is the intended audience?

**OBJECTIVITY**

Are advertising, opinion, and factual information clearly distinguished?

What biases, if any, can you detect?

**Accuracy**

Is important information documented through links so that it can be verified in other sources?
Is the text well written and free of careless errors in spelling and grammar?

**Coverage and currency**

Is there any indication that the site is still under construction?
For sources with print equivalents, is the Web version more or less extensive?
How detailed is the treatment of the topic?
Is there any indication of the currency of the information (date of last update or statement regarding frequency of updates)?
Are the graphics helpful, or are they simply window dressing?

## 48 Keeping a Working Bibliography

As you discover books, journal articles, newspaper stories, and Web sites that you think might be helpful, you need to maintain a record of vital information about each source. This record, called a working bibliography, will enable you to know where sources are as well as what they are when it comes time to consult them or to acknowledge them in your list of works cited or final bibliography (see pages 184–195, 207–213, 220–226, and 234–237). In all likelihood, your working bibliography will contain more sources than you eventually use and include in your list of works cited.

An easy way to develop a working bibliography is to record all bibliographic information on 3" × 5" index cards. Using separate cards for each book, article, Web site, or other online source allows you to edit your working bibliography continually, dropping sources that are not helpful for one reason or another and adding new ones. With the computerization of most library resources, you now have the option of printing out bibliographic information from the computer catalog and periodical indexes or from the Internet. Be sure to check the bibliographic information on your printouts against the Checklist for a Working Bibliography, and supply any information that is missing.

## Checklist for a Working Bibliography

**For books**

Library call number
Names of all authors, editors, and translators
Title and subtitle
Publication data:
  Place of publication (city and state)
  Publisher's name
  Date of publication
Edition (if not the first) and volume number (if applicable)

**For periodical articles**

Names of all authors
Title and subtitle of the article
Title of journal, magazine, or newspaper
Publication data:
  Volume number and issue number
  Date of issue
  Page numbers

**For Internet sources**

Names of all authors, editors, compilers, or sponsoring agents
Title and subtitle of the document
Title of the longer work to which the document belongs (if applicable)
Title of the site or discussion list name
Author, editor, or compiler of the Web site or online database
Date of release, online posting, or latest revision
Name and vendor of database or name of online service or network
Medium (online, CD-ROM, etc.)

Format of online source (Web page, e-mail, etc.)
Date you accessed the site
Electronic address: URL or network path

**For other sources**

Name of author, government agency, organization, company, recording artist, personality, and so forth
Title of the work
Format (pamphlet, unpublished diary, interview, television broadcast, etc.)
Publication or production data:
    Name of publisher or producer
    Date of publication, production, or release
    Identifying codes or numbers (if applicable)

## 49 Avoiding Plagiarism

The importance of honesty and accuracy in doing library research cannot be stressed enough. Any material borrowed word for word must be placed within quotation marks and properly cited; any idea, explanation, or argument you borrow in a summary or paraphrase must be documented, and it must be clear where the borrowed material begins and ends. In short, to use someone else's ideas in their original form or in an altered form without proper acknowledgment is plagiarism. Plagiarism is a serious offense regardless of whether it occurs intentionally or accidentally.

Your citations must consistently follow an established system of documentation, such as that of the Modern Language Association (see MLA Style, pages 178–195), the American Psychological Association (see APA Style, pages 203–213), the *Chicago Manual* (see pages 220–226), or the Council of Science Editors (see CSE style, pages 233–237). Your instructor will tell you which to use.

Do not clutter your paper with citations for facts or common knowledge. For example, it is common knowledge that Sandra Day O'Connor is the first woman to serve on the Supreme Court and that the Soviet Union ceased to exist in

1992. If you find the same information repeated in several sources, you can assume that it is common knowledge and that you do not need to document it. If, however, you find information that is already documented, it is a sure sign that you need to document it.

A little attention and effort at certain stages in the research process can go a long way in eliminating the possibility of inadvertent plagiarism. At the note-taking stage, check what you record against the original, paying particular attention to word choice and word order, especially if you are paraphrasing. It is not enough simply to use a synonym here or there and think it is a paraphrase; you must restate the idea in your own words, using your own style. At the drafting stage, check whenever you incorporate a source into your paper, be careful to put quotation marks around material taken verbatim, and double-check your text against your notes—or, better yet, against the original source if you have it on hand.

The following example illustrates how plagiarism can occur when care is not taken in the wording or sentence structure of a paraphrase. Finally, an acceptable paraphrase of the original source is given.

**ORIGINAL SOURCE**

In theory, recycled paper should be cheaper, since it turns garbage that someone would otherwise have to pay to dispose of into a marketable commodity.

—Leslie Pardue, "What Goes Around Comes Around," *E Magazine* Mar.–Apr. 1990: 56

**UNACCEPTABLY CLOSE WORDING**

➤ According to Leslie Pardue, turning someone's garbage into a marketable product while avoiding the costs of disposal would seem in theory to make recycled paper less expensive (56).

**UNACCEPTABLY CLOSE SENTENCE STRUCTURE**

➤ According to Leslie Pardue, it seems only reasonable that recycled paper would be less expensive because it turns someone's waste that would otherwise have to be discarded into a new product (56).

**ACCEPTABLE PARAPHRASE**

➤ By eliminating landfill costs, recycling, according to Leslie Pardue, should enable people to produce paper that is less expensive than paper made from virgin pulp (56).

If at any time while you are taking notes or writing your paper you have a question about plagiarism, it is your responsibility to consult your instructor for clarification and guidance before proceeding.

# 50 Integrating Quotations, Paraphrases, Summaries, and Visuals

When you refer to borrowed material in your research paper, whether by summary, paraphrase, or direct quotation, it is best to introduce the material with a signal phrase, which alerts the reader that the borrowed information is about to be stated.

## 50a Signal phrases

*Signal phrases* (e.g., *according to Leslie Pardue*) help the reader follow your train of thought. They also tell the reader who is speaking and, in the case of summaries and paraphrases, indicate exactly where your ideas end and someone else's begin. Without a signal phrase to integrate a quotation into the flow of your paper, the quotation seems to come from nowhere and jars the reader. Consider the following example in which the quoted material is simply dropped into the text:

UNANNOUNCED QUOTATION

➤ Most Americans think that we are producing more trash per person than ever, that plastic is a huge problem, and that paper biodegrades quickly in landfills. "The biggest challenge we will face is to recognize that the conventional wisdom about garbage is often wrong" (Rathje 99).

Now look at what happens when the writer uses an appropriate signal phrase to introduce the quotation:

INTEGRATED QUOTATION

➤ Most Americans think that we are producing more trash per person than ever, that plastic is a huge problem, and that paper biodegrades quickly in landfills. William L. Rathje, director of the Garbage Project at the

University of Arizona, views the issue differently; he believes that "the biggest challenge we will face is to recognize that the conventional wisdom about garbage is often wrong" (99).

The quotation in this example is integrated into the text not only by means of a signal phrase, but in two other ways as well. By mentioning the speaker's authority and that he "views the issue differently," the writer provides even more of a context for the reader to understand how the quotation fits into the discussion.

How well you integrate a quotation, paraphrase, or summary into your paper depends on your choice of signal phrase—in particular, the verb in the signal phrase. It is the verb that conveys the tone and intent of the writer being cited. If a source is arguing, use the verb *argue* (or *asserts, claims,* or *contends*); if a source contests a particular position or fact, use *disagrees* (or *denies, disputes, refutes,* or *rejects*); if a source agrees with a particular position or fact, use *concurs* (or *admits, concedes,* or *grants*). By using verbs that are specific to the situation in your paper, you help the reader process and organize the information you are presenting and thereby better understand your use of the material.

Other verbs that you should keep in mind when constructing signal phrases include the following:

| | | |
|---|---|---|
| acknowledges | declares | points out |
| adds | endorses | reasons |
| admits | grants | reports |
| believes | implies | responds |
| compares | insists | suggests |
| confirms | | |

**NOTE:** The American Psychological Association prefers that the past tense (e.g., *declared, reported*) or the present perfect tense (e.g., *has declared, have reported*) be used for verbs in signal phrases.

## 50b Quotations, paraphrases, and summaries

Using quotations, paraphrases, and summaries purposefully means using only those that truly support your thesis.

## Quotations

Using quotations effectively means reserving direct quotation for important ideas stated in vivid, imaginative, and memorable language or—when writing about literature—for revealing statements by characters themselves. When you are considering using a quotation, ask yourself these questions:

- How well does the quotation illustrate or support my analysis?
- Is this quotation the best evidence of the point I am making?
- Why am I quoting the text instead of paraphrasing or summarizing it?

Resist the temptation to use quotations simply because they sound good or because you think you do not have "enough" of them in your paper. Finally, remember that the appropriate use of quotations gives your paper authority.

Whenever you quote, you need to tell how the quotation relates to your point. Do not leave it to your readers to guess your intent. Introduce each quotation, and draw a conclusion from it. Integrate quotations into the flow of your paper so that they are not stylistically jarring. Consider the following uses of quotations in a student paper on the short stories of Edgar Allan Poe.

**WEAK**

➤ Poe often uses specific details from the settings of his stories to suggest the psychological states of his characters. For example, in describing the "barely perceptible fissure" (237), he prefigures Roderick's mental disintegration.

**EFFECTIVE**

➤ Poe often uses specific details from the settings of his stories to suggest the psychological states of his characters. In "The Fall of the House of Usher," for example, the narrator's description of the "barely perceptible fissure" (237) in the front wall of the Usher mansion prefigures Roderick's own mental disintegration.

Naming the short story and identifying the words as those of the narrator provide more of a context for the quotation. Stating that the fissure was in the front wall of the Usher mansion makes explicit the comparison between setting and psychological state.

Paraphrases

When you paraphrase a source, you restate the information in your own words instead of quoting the source directly. A paraphrase presents the information in the original in approximately the same number of words, but the wording is your own. That is, a paraphrase should closely parallel the ideas presented in the original but not use the same words or sentence structure as the original. To do so would be to plagiarize.

The following is original source material that a student found in her research:

> Curbside recycling did not begin in earnest until the mid-1980s, when the national recycling rate was 3 percent. Today the rate hovers around 22 percent. Some 3,700 U.S. curbside collection programs existed in 1993 vs. 600 five years before. Some of these programs are expensive turkeys; most are affordable and successful. As a result, even with ever-increasing consumption in several major categories of consumer products, the amount of materials sent to municipal landfills peaked in the late 1980s and has been declining since. Already, several states and cities have recycling levels as high as those found in Japan. New Jersey has gone from a standing start to recycling 53 percent of municipal wastes, under a program whose target was 25 percent. Seattle recycles about 50 percent, the level achieved in Tokyo.
>
> —Gregg Easterbrook, "Good News from Planet Earth," *USA Weekend* 14–16 Apr. 1995: 5.

The student then paraphrased the source and used it in her paper as follows:

➤ According to Easterbrook, recycling programs have had a tremendous impact on the problems of waste disposal in this country. Since its inception in the mid-1980s, curbside recycling has risen from 3 percent to almost 22 percent today. And in the years 1988 through 1993, curbside collection programs increased more than sixfold. Not coincidentally,

> Americans have managed to keep landfill deposits down while their rate of consumption continues to rise. The state of New Jersey and the city of Seattle have model recycling programs, rivaling the results already achieved in Japan (5).

The signal phrase at the beginning and the parenthetical page citation at the end clearly indicate what is being borrowed. The absence of quotation marks indicated that the material is in the student's own words.

### Summaries

When you summarize material from a source, you capture in your own words the essential idea of a passage or of an entire chapter or article in highly condensed form. Summaries are particularly useful when you are working with lengthy, detailed arguments or long passages of descriptive background information. You can distill a chapter or more into a paragraph or several paragraphs into a sentence or two. One student took a detailed two-paragraph analogy between an iceberg and waste or trash from a book and rendered it into the following summary, which appeared in his paper:

➤ In The Waste Watchers, A. H. Purcell likens waste to an iceberg. Because we are not aware of the size/amount of our waste, the danger is greater than we think (14–15).

The student's summary captures the spirit of the analogy while foregoing the details of it, and again the signal phrase and the parenthetical page citation mark the beginning and end of the summary.

## 50c Visuals

Visuals—tables, charts, graphs, diagrams, illustrations, photographs, and clip art—can add dimension and substance to your papers. They catch attention, illustrate concepts, carry meaning, and allow readers to see as well as read the information you are presenting. Well-chosen visuals support and clarify information already presented in writing.

As with quotations, paraphrases, and summaries, you should integrate visuals into the text of your paper with a signal phrase or a sentence or two that sets the context. Consider the following example:

➤ In its annual report for 2000, VeloPazzo Software announced that overall sales increased for the eastern and central regions and declined for the western region:

LINE GRAPH

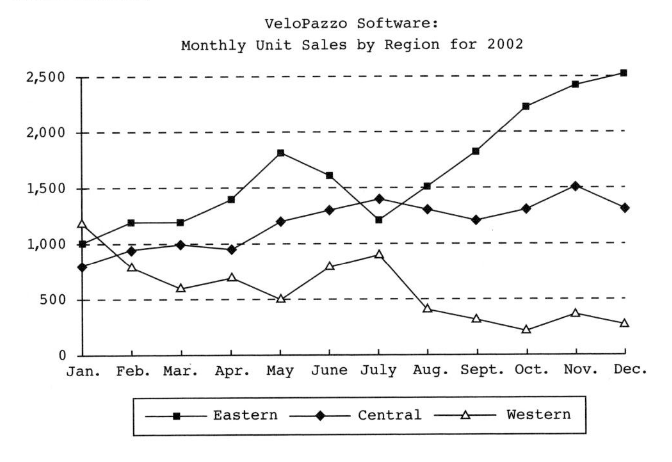


## 50d Long quotations

Set off, or block, prose quotations that are longer than four lines to help your reader more clearly see the quotation as a whole. Set off verse quotations longer than three lines. Set-off quotations are indented one inch (or ten spaces) from the left margin and double-spaced; no quotation marks are necessary because the format itself indicates the passage is a quotation. When you are quoting two or more paragraphs from the same source, indent the first line of each paragraph an additional quarter inch (or three spaces).

The following example is taken from a student's library research paper on John Updike's story "A Sense of Shelter":

➤ Fearing rejection from his fellow high school seniors, William chooses to sit with the more accepting juniors, who tend to look up to him as an older student. In addition, Updike sharply contrasts William's behavior with that of the character he loves, Mary, who is more adventurous. Critic A. S. G. Edwards explains this contrast:

> On each of the three occasions she appears, Updike presents her going away from William's world, away from the classroom, out of the soda bar, and finally out of the school itself. This final encounter effectively dramatizes the nature of the contrasting characters: Mary's steady progression through and finally away from the school out into the cold world more appropriate to her greater experience and maturity is set against William's hurried retreat into the womb-like warmth of the school. (467)

The colon following the student's sentence introducing the quotation indicates that the quotation is closely related to what precedes it. Note that, unlike an integrated quotation in which the parenthetical citation is inside the end punctuation, with a long quotation the parenthetical citation goes outside the final punctuation.

On occasion, long quotations contain more information than you need to use. In such situations you can use an ellipsis mark (pages 112–113) to show that you omitted material from the original text, in this way keeping the quotation to the point and as short as possible.

Another mark of punctuation that is useful when working with quotations is brackets (see page 111). These squared parentheses allow you to insert your own words into a quotation to explain a pronoun reference, for example, or to change a verb tense to fit your text.

➤ According to Rathje, "There are no ways of dealing with it [garbage] that haven't been known for many thousands of years" (100).

## 51 Revising the Research Paper

A good research paper is the product of a sustained research and writing effort that can stretch over a period of four weeks or more. Revision is one of the key steps in the process because it offers you the opportunity to change and rewrite your paper so that it says what you want it to say. As with any paper, start by looking at the large elements of thesis, purpose, organization, paragraph development, and supporting evidence. Use the checklist on page 21 to guide your handling of these large elements. Later, you can turn to matters of style, diction, and correctness (see pages 21–22). With a research paper, you need to pay particular attention to your use of source material, making sure that each citation is accurately represented and that its purpose in the context of your paper is clear. Use the Checklist for Revising a Research Paper as you reread your initial draft, looking for opportunities to make the main point of your paper clearer, the content of your paper stronger, the evidence in your paper more compelling, and the style in your paper more readable.

## Checklist for Revising a Research Paper

**QUESTIONS TO ASK ABOUT CONTENT, ORGANIZATION, AND STYLE**

1. Is your thesis stated clearly, and does it accurately reflect your stand on the topic? Is it positioned so that readers cannot miss it? (See pages 12–13.)
2. Is your purpose in writing the research paper clear? (See page 9.)
3. Is the paper unified and coherent? That is, does each of your paragraphs relate to or support the thesis?
4. Is the organization of your paper easy to follow? Is there a logic to the order in which ideas are presented? (See pages 16–18.)
5. Are your important points given adequate emphasis?
6. Are transitions present to assist readers from one paragraph or group of paragraphs to the next? (See pages 19–20.)
7. Do your ideas and not those of your sources control the paper? Have you missed opportunities to synthesize ideas or draw conclusions based on material found in your sources?
8. Is your argument persuasive on all points, or are there places where you could use more evidence? (See pages 43–44.)

9. Have you accounted for opposing arguments? (See page 42.)
10. Have you integrated your sources into your paper smoothly? (See pages 167–173.)
11. Are your sentences clear and varied? (See pages 33–36.)
12. Is your voice appropriate, given your subject and your purpose? (See pages 37–40.)

**QUESTIONS TO ASK ABOUT USE OF SOURCES**

*Quotations*

1. Do quotation marks clearly indicate the language that you borrowed verbatim? (See pages 104–105.)
2. Is the language of the quotation accurate? (See pages 165–167.)
3. Do brackets or ellipsis marks clearly indicate changes or omissions you have introduced? (See pages 111–113.)
4. Does a signal phrase naming the author introduce each quotation? Does the verb in the signal phrase help to establish a context for each quotation? (See pages 167–168.)
5. Does a parenthetical citation follow each quotation? (See pages 178–184.)

*Summaries and paraphrases*

1. Are all summaries and paraphrases written in your own words and style so as not to constitute plagiarism? (See pages 165–167 and 168–171.)
2. Do all summaries and paraphrases accurately represent the opinion, position, or reasoning of the original source? (See pages 170–171.)
3. Do all summaries and paraphrases start with a signal phrase so that readers know where the borrowed material begins? (See pages 167–171 and 178–184.)
4. Do all summaries and paraphrases conclude with a proper parenthetical citation? (See pages 178–184.)

*Facts and statistics*

1. Is each fact or statistic that is not common knowledge clearly documented with a parenthetical citation? (See pages 165–167.)

# *Documentation*

In writing a research paper, you are using the information and ideas of others. Whenever you directly quote, summarize, or paraphrase another person's thoughts and ideas or use facts and statistics that are not commonly known, you must document, or properly acknowledge, your source.

# 52 MLA Documentation Style

The documentation system recommended by the Modern Language Association (MLA) in the *MLA Handbook for Writers of Research Papers,* fifth edition (1999), is discussed and illustrated in this section. The MLA system is used primarily, although not exclusively, in English, foreign languages, and the humanities. We also discuss the Alliance for Computers and Writing (ACW) guidelines for citing electronic sources.

## 52a MLA in-text citations

In-text citations are easy to use and informative because they immediately let readers know the source of a citation without breaking the flow of the paper. Complete publication data are provided at the end of the paper in a Works Cited list.

Most MLA-style in-text citations require only the author's last name and a page reference. The customary way to integrate this information into your text is to give an introductory signal phrase mentioning the author's name at the beginning of the sentence and then the page number parenthetically at the end of the borrowed material. If the author's name is not given in the signal phrase, it must appear parenthetically with the page number.

The following examples show correct MLA in-text citation form for different kinds of sources. Note that the end punctuation comes after the parenthetical citation, no *p.* for *page* is used, and there is no punctuation between author and page number.

## Directory to MLA In-Text Citations



#### 1. AUTHOR MENTIONED IN A SIGNAL PHRASE

Rich argues that the Napoleonic Wars were "world wars in the sense that they involved all the great powers of that era" (18).

#### 2. AUTHOR NOT MENTIONED IN A SIGNAL PHRASE

As one authority in diplomatic history argues, the Napoleonic Wars were "world wars in the sense that they involved all the great powers of that era" (Rich 18).

#### 3. TWO OR MORE WORKS BY THE SAME AUTHOR

You can handle this in one of three ways. Give the title of the relevant work in the signal phrase.

In "Politics and the English Language," Orwell claims that "political language is designed to make lies sound truthful and murder respectable" (508).

Or give the title parenthetically at the end in an abbreviated form.

Orwell claims that "political language is designed to make lies sound truthful and murder respectable" ("Politics" 508).

In the absence of a signal phrase, put the author's name, the abbreviated title, and the page reference in parentheses.

After killing the elephant, "I often wondered whether any of the others grasped that I had done it solely to avoid looking a fool" (Orwell, "Shooting" 46-47).

**4. TWO OR THREE AUTHORS**

Miller and Swift argue that "masculine and feminine are as sexist as any words can be because they invoke cultural stereotypes" (176).

Probability theory is a prerequisite for advanced work in the social sciences at colleges and universities today (Hoel, Port, and Stone 27-54).

**5. MORE THAN THREE AUTHORS**

For works with more than three authors, include only the last name of the first author followed by the Latin phrase *et al.* (meaning "and others").

"As a whole Gulliver's Travels has the multiple intentions of a masterpiece" (Baugh et al. 865).

**6. AUTHORS WITH THE SAME LAST NAME**

To avoid confusion when using two or more authors with the same last name, give each author's first name and last name in all references.

Donald H. Graves believes successful teachers of writing have firm control of the twin crafts, teaching and writing (5-6).

Other educators have developed ways in which teachers can transfer this control to their students (Richard L. Graves 239).

**7. CORPORATE AUTHOR**

A corporate author is an institution, company, agency, or organization that is credited with authorship of a work. Treat the name just as you would an individual, giving it in the signal phrase or in parentheses. The first time you use a corporate name, spell it out (National Organization for Women).

However, if it has a commonly used abbreviation (NOW), give the abbreviation in parentheses immediately after the first reference to the organization. Thereafter you can simply use the abbreviation in the text.

> The National Rifle Association (NRA) opposes legislation--including the historic Brady Bill--that seeks to restrict the sale of handguns (4-17).

**8. UNKNOWN AUTHOR**

Use the complete title of the work in a signal phrase or a shortened version in the parenthetical reference.

> According to the article "It's Raining Pennies--But Not from Heaven," the penny "is under attack as a public nuisance" (31).

> The penny "is under attack as a public nuisance" ("It's Raining Pennies" 31).

**9. MULTIVOLUME WORK**

Include the volume number followed by a colon and then the page number in your parenthetical citations.

> Manchel has little doubt that "the basic metaphorical images of the American Dream grew out of seventeenth-century European Utopian visions" (2: 1046).

**10. AN ENTIRE WORK**

To cite a whole book or article, give the author's name in the signal phrase or parenthetically; no page numbers are needed.

> Peggy Noonan argues for the historical importance of political speeches.

> The historical importance of political speeches should not be overlooked (Noonan).

**11. FICTION, POETRY, DRAMA, AND THE BIBLE**

Because the Bible and literary works are often available in different editions, you need to include information that will help your readers locate the particular reference in the edition that they have access to.

#### FICTION

Give the page number of the edition you are using followed by a semicolon, and then give the part or chapter number.

> Edna Pontellier, the heroine of Chopin's The Awakening, stands alone in triumph at the novel's end. In the words of the narrator, "How strange and awful it seemed to stand naked under the sky!" (301; ch. 39).

#### POETRY

First give the part or division reference, if there is one, and then the line numbers.

> In "Song of Myself," Whitman emerges larger than life itself--"a kosmos, of Manhattan the son, / Turbulent, fleshy, sensual, eating, drinking, and breeding" (24.1-2).

#### DRAMA

Include the act, scene, and lines in your citation. Use periods to separate these items.

> Shakespeare's King Lear projects his own misery when he asks the Fool, "What, ha's his own daughters brought him to this pass? / Couldst thou save nothing? Didst thou give them all?" (3.4.61-62).

#### BIBLE

In citing the Bible, provide book, chapter, and verse. As with drama citations, use periods to separate these items. The books of the Bible may be abbreviated in the parenthetical citation but not when part of your text.

> In the gospel according to St. Matthew, Christ begins his Sermon on the Mount by declaring who are the blessed (5.3-12).

> Later in his sermon, Christ gives his listeners what we now know as the Lord's Prayer (Matt. 6.9-13).

#### 12. WORK IN AN ANTHOLOGY

Give the name of the writer of the piece you are using—not the editor of the anthology—in your signal phrase or parenthetical citation. See number 14 on page 187.

> Dao uses striking figures of speech in the second half of the story (287–92).

#### 13. A SOURCE QUOTED IN ANOTHER SOURCE

It is always better to take your material from original sources when they are available. Sometimes only indirect sources are available. If you need to cite someone quoted in a work written by another person, begin your parenthetical citation with *qtd. in* (for "quoted in").

> To Boris Yeltsin, "Gorbachev was a communist, is a communist, and will always be a communist" (qtd. in Edwards 15).

#### 14. TWO OR MORE SOURCES IN A SINGLE CITATION

When acknowledging more than one source in a single parenthetical citation, separate the citations with a semicolon.

> Environmentalists contend that solid waste disposal will be America's--if not the world's--most pressing problem in the next decade (Blumberg and Gottlieb 117; Rathje 99).

#### 15. INTERVIEW

To cite a personal interview that you conducted, simply give the person's name in a signal phrase or parenthetically.

> The folklorist Jan Brunvald discussed how difficult it is to trace the origins of these popular, widely circulated stories.

> The origins of urban legends are extremely difficult to trace (Brunvald).

#### 16. AN ELECTRONIC SOURCE

Cite an electronic source much as you would a print source. If you know both the author's name and the page number, present both.

> Using Gallup poll results from the last twenty years, Mark Gillespie points out that "critics of capital punishment contend it unfairly targets minorities and the poor, and the American public tends to agree" (2).

Often electronic sources do not have page numbers. Some electronic sources use alternative numbering systems such as paragraphs, sections, or screens. Use the abbreviations *par.* and *sec.* or the word *screen* together with the number in such cases.

> Advances in cloning research offer hope to patients with chronic illnesses or terminal diseases (Meagher, par. 9).

When there is no author identified, give the complete title in your signal phrase or a shortened title in parentheses.

> According to the Web page sponsored by the Ohio Newspaper Association, the recycled content of newsprint in Ohio was 36.3 percent in 1999 ("Recycled").

## 52b MLA list of works cited

The MLA-style Works Cited section is an alphabetical listing of the complete publication data for each source you cite in your paper. It is not a list of all the sources you checked, just the ones you actually cite.

### Directory to MLA List of Works Cited

#### Books



#### Periodicals

**Electronic sources**



**Other sources**

## Books

**1. ONE AUTHOR**

Croft, Robert W. Anne Tyler: A Bio-Bibliography. Westport: Greenwood, 1995.

**2. TWO OR THREE AUTHORS**

Young Bear, Severt, and R. D. Theisz. Standing in the Light: A Lakota Way of Seeing. Lincoln: U of Nebraska P, 1994.

The name of the first author is the only one reversed; separate the names with commas.

Brunas, Michael, John Brunas, and Tom Weaver. Universal Horrors: The Studio's Classic Films, 1931-1946. Jefferson: McFarland, 1990.

**3. MORE THAN THREE AUTHORS**

Bloom, Harold, et al. Deconstruction and Criticism. New York: Continuum, 1979.

**4. AUTHOR WITH AN EDITOR**

Fisher, Dorothy Canfield. The Bedquilt and Other Stories. Ed. Mark J. Madigan. Columbia: U of Missouri P, 1996.

**5. EDITORS**

Gwertzman, Bernard, and Michael T. Kaufman, eds. The Collapse of Communism. New York: Random, 1990.

**6. TRANSLATION**

Kafka, Franz. The Penal Colony: Stories and Short Pieces. Trans. Willa Muir and Edwin Muir. New York: Schocken, 1984.

**7. CORPORATE AUTHOR**

English Language Center, Brigham Young University. Expeditions into English: Grammar 1. Englewood Cliffs: Prentice, 1991.

**8. UNKNOWN AUTHOR**

Children of the Dragon: The Story of Tiananmen Square. New York: Collier-Macmillan, 1990.

9. TWO OR MORE WORKS BY THE SAME AUTHOR

Dillard, Annie. An American Childhood. New York: Harper, 1987.

---. The Living: A Novel. New York: HarperCollins, 1992.

10. EDITION OTHER THAN FIRST

Aitchison, Jean. Language Change: Process or Decay? 2nd ed. Cambridge: Cambridge UP, 1991.

11. REPUBLISHED BOOK

Wharton, Edith. The House of Mirth. 1905. Introd. Cynthia Griffin Wolff. New York: Penguin, 1985.

12. WORK IN MORE THAN ONE VOLUME

Manchel, Frank. Film Study: An Analytical Bibliography. 4 vols. Rutherford: Fairleigh Dickinson UP, 1990-91.

Give the number of volumes after the title, using the abbreviation *vols.* This multivolume work was published over a two-year period. If you use only a particular volume, name it specifically before the city and publisher; although not required, the total number of volumes may be added at the end of the entry as follows:

Manchel, Frank. Film Study: An Analytical Bibliography. Vol. 2. Rutherford: Fairleigh Dickinson UP, 1990-91. 4 vols.

13. WORK IN A SERIES

Magistrale, Tony. Stephen King: The Second Decade, Danse Macabre to The Dark Half. Twayne's United States Authors Series: 599. New York: Maxwell Macmillan International, 1992.

14. WORK IN AN ANTHOLOGY

First give the title of the work, then the title and editor(s) of the anthology. Cite the pages on which the work appears after the publication data.

Dao, Bei. "13 Happiness Street." Contemporary Literature of Asia. Ed. Arthur W. Biddle, Gloria Bien, and Vinay Dharwadker. Upper Saddle River: Prentice, 1996. 281-92.

#### 15. TWO OR MORE WORKS FROM THE SAME ANTHOLOGY

When using two or more works from the same anthology, you can avoid repetition by first citing the anthology with complete publication data.

Shain, Charles, and Samuella Shain, eds. The Maine Reader: The Down East Experience 1614 to the Present. Boston: Houghton, 1991.

The entries for the selections you used from the anthology need then include only the author's name, the title of the selection, the editors' name, and the page numbers.

Chute, Carolyn. "Earlene." Shain and Shain 506-13.

Thoreau, Henry David. "A Moose Hunt." Shain and Shain 138-48.

White, E. B. "Home-Coming." Shain and Shain 395-402.

#### 16. PREFACE, INTRODUCTION, FOREWORD, OR AFTERWORD

Douglas, Ann. Introduction. Uncle Tom's Cabin or, Life among the Lowly. By Harriet Beecher Stowe. New York: Penguin, 1981. 7-36.

#### 17. ENCYCLOPEDIAS AND OTHER REFERENCE WORKS

Preuss, Harry G. "Nutrition." The Encyclopedia Americana. 1994 ed.

### Periodicals

#### 18. ARTICLE IN A MONTHLY MAGAZINE

Norman, Geraldine. "Russia's Rocky Past." Arts and Antiques Oct. 1997: 94-99.

#### 19. ARTICLE IN A WEEKLY MAGAZINE

McLeod, Scott. "The Nile's Other Kingdom." Time 15 Sept. 1997: 102-06.

#### 20. ARTICLE IN A JOURNAL PAGINATED BY VOLUME

Flower, Dean. "Desegregating the Syllabus." Hudson Review 46 (1994): 683-93.

#### 21. ARTICLE IN A JOURNAL PAGINATED BY ISSUE

Deegan, James G. "Gender and Pedagogy in Soviet Education: The Inherited Context." International Education 22.2 (1993): 46-69.

**22. ARTICLE IN A DAILY NEWSPAPER**

Broad, William J. "Another Possible Climate Culprit: The Sun." New York Times 23 Sept. 1997, New England final ed.: C1+.

**23. ARTICLE IN A WEEKLY NEWSPAPER**

Lieb, Kristin. "Minority Students Get Help in Seeking Advanced Degrees." Chronicle of Higher Education 5 Aug. 1992: A27-28.

**24. EDITORIALS AND LETTERS TO THE EDITOR**

Diem, Bui. Letter. Washington Post 11 May 1995, late ed.: A22.

Gough, Patricia B. "Moving beyond the Myth." Editorial. Phi Delta Kappan 76 (1994): 99.

**25. BOOK REVIEW**

Angell, Roger. "Designated Hero." Rev. of Jackie Robinson, by Arnold Rampersad. New Yorker 15 Sept. 1997: 82-88.

**26. FILM REVIEW**

Holden, Stephen. "A Love Affair Kept All in the Family." Rev. of The House of Yes, dir. Mark Waters. New York Times 10 Oct. 1997, New England final ed.: B10.

## Electronic sources

The following guidelines for citing information retrieved from the World Wide Web are those provided on the MLA's Web site <http://www.mla.org>. All of the following elements, if they are available and relevant, should be included in the reference. All items should be followed by a period with the exception of the date the site is accessed.

- *Name of the author,* editor, compiler, or translator of the source (if available and relevant), reversed for alphabetizing and followed by an abbreviation, such as *ed.,* if appropriate.
- *Title of a poem,* short story, article, or similar short work within a scholarly project, database, or periodical (in quotation marks); or title of a posting to a discussion list or forum (taken from the subject line and put in quotation marks), followed by the description *Online posting.*

- *Title of a book* (underlined).
- *Name of the editor,* compiler, or translator of the text (if relevant and if not cited earlier), preceded by the appropriate abbreviation, such as *Ed.*
- *Publication information* for any print version of the source.
- *Title of the scholarly project,* database, periodical, or professional or personal site (underlined); or, for professional or personal site with no title, a description such as *Home page.*
- *Name of the editor* of the scholarly project or database (if available).
- *Version number of the source* (if not part of the title) or, for a journal, the volume number, issue number, or other identifying number.
- *Date of electronic publication,* of the latest update, or of posting.
- *For a work from a subscription service,* the name of the service and—if a library is the subscriber—the name and city (and state abbreviation, if necessary) of the library.
- *For a posting to a discussion list or forum, the name of the list or forum.*
- *The number range or total number of pages,* paragraphs, or other sections, if they are numbered.
- *Name of any institution or organization* sponsoring or associated with the Web site.
- *Date when the researcher accessed the source.*
- *Electronic address,* or URL, of the source (in angle brackets); or, for a subscription service, the URL of the service's main page (if known) or the keyword assigned by the service.

Your instructor may recommend that you follow guidelines set forth in *The Columbia Online Style,* developed by Janice R. Walker and Todd Taylor of the University of South Florida.

The Alliance for Computers and Writing (ACW) has endorsed COS style. For current recommendations as well as new developments in COS style, visit the Columbia University Press Web site at <http://www.columbia.edu/cu/cup/cgos/idx_basic.html>.

#### 27. A SOURCE ON A PERIODICAL CD-ROM

Some CD-ROM databases are updated on a regular basis because they cover publications that are themselves published periodically. Your entry in the Works Cited list should include the publication information for the print source, followed by the title of the database (underlined), the publication medium (CD-ROM), the name of the vendor, and the electronic publication date.

> James, Caryn. "An Army Family as Strong as Its Weakest Link." New York Times 16 Sept. 1994, late ed.; C8. New York Times Ondisc. CD-ROM. UMI-Proquest. Oct. 1994.

#### 28. A SOURCE ON A NONPERIODICAL CD-ROM

Some CD-ROMs are published as a single edition much as books are. Cite these electronic sources as you would a book, being careful to add *CD-ROM* as the publication medium.

> Shakespeare, William. Macbeth. Ed. A. R. Branmuller. CD-ROM. New York: Voyager, 1994.

> "Proactive." The Oxford English Dictionary. 2nd ed. CD-ROM. Oxford, Eng.: Oxford UP, 1992.

#### 29. ONLINE BOOK

For a book available online, provide the author, the title, the editor (if any), publication information, the date of access, and the complete electronic address.

> Hawthorne, Nathaniel. Twice-Told Tales. Ed. George Parsons Lathrop. Boston: Houghton, 1883. 1 Mar. 2001 <http://www.tiac.net/users/eldred/nh/ttt.html>.

For an online book that is part of a scholarly project, include information about the project immediately after the information about the printed book.

> Butler, Josephine E. The Education and Employment of Women. Liverpool, 1868. Victorian Women Writers Project. Ed. Perry Willett. June 1996. Indiana U. 4 Feb. 1998 <http://www.indiana.edu/~letrs/vwwp/butler/educ.html>.

**30. ARTICLE IN AN ONLINE JOURNAL**

Provide the author, the title of the article, the title of the journal, the volume and issue, and the date of issue, followed by the number of paragraphs in the article, the date of access, and the electronic address.

Johnson, Diane. "Helping Children Succeed after Divorce: Building a Community-Based Program in a Rural County." Journal of Extension 38.5 (Oct. 2000): 17 pars. 30 Jan. 2001 <http://joe.org/joe/2000october/iwl.html>.

**31. ARTICLE IN AN ONLINE NEWSPAPER**

Elliot, Stuart. "Nielsen Unit Offers Data about Internet Users." New York Times on the Web 22 July 1996. 2 Feb. 2001 <http://www.nytimes.com/ library/articles/users.html>.

**32. ARTICLE IN AN ONLINE MAGAZINE**

Holloway, Marguerite. "Pattie." Wired 5.12. Dec. 1997, 2 Feb. 2001 <http://www.wired.com/wired/5.12/maes.html>.

**33. ONLINE SCHOLARLY PROJECT OR DATABASE**

Victorian Women Writers Project. Ed. Perry Willett. June 1996. Indiana U. 27 July 2000 <http://www.indiana.edu/~letrs/vwwp/>.

Electrifying the Renaissance: Hypertext, Literature, and the World Wide Web. Mar. 1997. U of California, Santa Barbara. 20 Apr. 2001 <http://humanitas.ucsb.edu/depts/english/english/coursework/rar>.

**34. ARTICLE IN AN ONLINE REFERENCE WORK**

"e. e. cummings." Encyclopedia.com. 1994. The Concise Columbia Electronic Encyclopedia. 3rd ed. 12 Sept. 1999 <wysiwyg://Body.30/ http://www.encyclopedia.com/articles/03328.htm>.

**35. PERSONAL ONLINE SITE**

Provide the author's name, the title, the access date, and the electronic address. In the absence of a title, use a label such as *Home page* to describe the site.

Cavrak, Steve. Home page. 15 Jan. 2001 <http://www.uvm.edu/~sjc/>.

#### 36. PROFESSIONAL ONLINE SITE

MLA on the Web. 20 Sept. 2000. Modern Language Association of America. 22 Jan. 2001 <http://www.mla.org>.

#### 37. A WORK FROM AN ONLINE SUBSCRIPTION SERVICE

When you access research materials through an online subscription service such as BRS, DIALOG, America Online, or Nexis, you should include the following information in your Works Cited entry: author's name, title and publication data for print source, title of the database (underlined), publication medium (*Online*), name of the computer service, and date you accessed the file. If an access or order number is available, include it.

Kappel-Smith, Diana. "Fickle Desert Blooms: Opulent One Year, No-Shows the Next." Smithsonian Magazine Mar. 1995: 9. America Online. 18 Apr. 1998. Keyword: desert.

#### 38. ELECTRONIC MAIL

Provide the writer's name, the title from the e-mail's subject heading, the word *E-mail* followed by the recipient, and the date of sending.

Quigley-Harris, Jennifer. "Re: Composition Classes." E-mail to author. 12 May 2001.

#### 39. POSTING TO A DISCUSSION LIST

Provide the writer's name, the title from the subject line, the words *Online posting,* the date of posting, the name of the forum, the date of access, and the electronic address.

Preston, Dennis R. "Re: Basketball Terms." Online posting. 8 Nov. 1997. American Dialect Society. 1 Mar. 1998 <http://www2.et.byu.edu/~lilliek/ads/indes.htm>.

#### 40. SYNCHRONOUS COMMUNICATION

When citing a synchronous communication from a MUD (multi-user domain) or a MOO (multi-user domain, object-oriented), provide the speaker's name, a description of the event, the date of the event, the forum, the date of access, and the electronic address.

Jensen, Carlos. “MediaMOO Symposium: Using 3D Graphics in Online Communities.” 20 Mar. 2000. MediaMOO. 2 Feb. 2001 <http://www.cc.gatech.edu/~carlosj/MediaMOO.html>.

#### 41. OTHER ONLINE SOURCES

When you use other information or materials accessed online, such as advertisements, audio recordings, charts, cartoons, maps, video clips, and works of art, cite them as you would if they had not been obtained online, adding the access date and the electronic address.

“San Francisco, CA.” Map. Yahoo! Maps. Yahoo! 2001. 10 Feb. 2001 <http://maps.yahoo.com>.

Seal, David. “Maps of the Solar System.” Photographs. SpaceArt. 1997. 1 Feb. 2001 <http://maps.jpl.nasa.gov>.

### Other sources

#### 42. PAMPHLET

NRA Institute for Legislative Action. The Myth of the “Saturday Night Special.” Washington: NRA, 1989.

#### 43. GOVERNMENT PUBLICATION

United States. Government Printing Office. Style Manual. Washington: GPO, 1984.

#### 44. PUBLISHED CONFERENCE PROCEEDINGS

McCloud, Gavin, ed. Restoring Sunken Vessels. Proc. of the Maritime and Shipping Industry’s Annual Conference on Ship Building, May 1990, U of Maryland. Annapolis: Annapolis P, 1991.

#### 45. UNPUBLISHED DISSERTATION

Taranto, Cheryl T. “Political Songsters for the Presidential Campaign of 1860.” Diss. Louisiana State U, 1994.

#### 46. LECTURE OR PUBLIC PRESENTATION

Bennett, Kristin. “Bottled Water: The Business of the 1990s.” Business and Management Lecture Series. Merrimack Coll. 14 Feb. 1998.

#### 47. LETTER

Proulx, E. Annie. Letter to the author. 22 Jan. 1998.

#### 48. LEGAL REFERENCE

R.A.V., Petitioner v. City of St. Paul, Minnesota. US Law Week. 60 LW 4667. US Supreme Ct. 1992.

#### 49. INTERVIEW

Gore, Tipper. Personal interview. 23 July 2000.

#### 50. FILM OR VIDEOTAPE

The Taming of the Shrew. Dir. Franco Zeffirelli. Perf. Richard Burton and Elizabeth Taylor. Columbia, 1966.

An Oral Historian's Work. With Edward D. Ives. Videocassette. Northeast Archives of Folklore and Oral History, 1987.

#### 51. TELEVISION OR RADIO PROGRAM

"Shakespearean Putdowns." Narr. Robert Siegel and Linda Wertheimer. All Things Considered. Natl. Public Radio. WBUR, Boston. 6 Apr. 1994.

#### 52. STAGE PLAY, OPERA, OR CONCERT PERFORMANCE

Il Trovatore. By Giuseppe Verdi. Cond. Simone Young. Perf. June Anderson. Metropolitan Opera House, New York. 17 Feb. 1998.

#### 53. WORK OF ART

Botticelli, Sandro. Birth of Venus. Uffizi Gallery, Florence.

#### 54. RECORD, TAPE, OR CD

Braxton, Toni. Secrets. LaFace, 1996.

#### 55. CARTOON

Koren, Edward. Cartoon. New Yorker 3 Oct. 1994: 54.

#### 56. MAP OR CHART

Philadelphia. Map. New York: Gousha, 1994.

#### 57. ADVERTISEMENT

Bose. Advertisement. The New York Times Magazine, 11 Nov. 2001: 129.

## 52c MLA information notes

Information notes come in two forms: content notes and bibliographic notes. Content notes allow the researcher to give readers additional information without digressing from the point at hand, whereas bibliographic notes provide helpful sources for additional reading not cited in the paper itself. For bibliographic notes, you must provide complete publication data as these works do not appear in the list of Works Cited. Information notes are optional in a research project. They may be presented at the foot of the page on which they appear or on a separate page at the end of the paper but before the Works Cited page. Indicate an information note by placing a raised (superscript) arabic numeral at the end of the sentence or passage to which the note refers. Number notes consecutively throughout the paper.

TEXT

Instead of entering the recycling loop--the cycle that takes a product from the consumer to the recycling bin to a manufacturer and back again to the consumer for use[1]--most old newsprint (ONP) sits in storehouses and trailer trucks with no place to go.

CONTENT INFORMATION NOTE

[1]For a thorough explanation of the processes involved in recycling paper, see Hagerty, Pavoni, and Heer (58-62) and Pardue (54-55). Shaw provides a discussion of the term "recycling loop" (1, 6).

## 52d MLA manuscript format

The following guidelines for formatting manuscripts have been adapted from Modern Language Association recommendations.

### Paper and type

For academic papers use 8 1/2" × 11", twenty-pound white paper, and type or print in black on one side of each sheet. Most instructors ask you to use a standard type style such as Times Roman or Courier, preferring that you avoid italic or script type styles for your main text. A letter-quality printer, such as a laser printer, gives the cleanest printout. If you use a dot matrix printer, be sure to use its letter-quality setting and a fresh ribbon. If you use continuous-feed paper, remove the perforated strips from the sides of the paper, separate the pages, and use a paper clip (do not staple) to secure the pages. Finally, be sure you keep both a paper copy and an electronic copy of your paper.

### Title page, title, name, and course information

Although MLA recommends omitting the title page, your instructor may require one. If so, place the title fifteen lines from the top, and center it. Approximately six lines below the title, center the word *by*, skip two more lines, and center your name. About twenty lines below your name, place the course title, professor's name, and date, all centered and double-spaced. Do not number the title page. See the sample on page 200.

Do not underline or italicize, put quotation marks around, or put a period at the end of the title of your paper. Capitalize the first letter of all words in the title except articles, coordinating conjunctions, and prepositions, unless they are the first or last word in a title.

If your instructor does not require a separate title page, type your name, your instructor's name, the name and number of the course, and the date on separate lines, double-spaced, beginning at the left margin one inch from the top of the first page. Double-space again, and center your title. Double-space between your title and the first sentence.

### Margins, spacing, and indentation

Leave a one-inch margin on all sides of the page. Some instructors prefer a one-and-one-half-inch margin on the left. Double-space the text of the paper, including set-off quotations, information notes, and the entries on the Works Cited page. Do not justify (make even) the right-hand margin. Indent the first line of each paragraph one-half inch (or five spaces).

For quotations of more than four typed lines of prose or three lines of poetry, indent one inch (or ten spaces) from the left margin. Double-space before, after, and within the quotation. Do not use quotation marks.

### Spacing for punctuation

Leave one space after a comma, colon, or semicolon and between the three periods in an ellipsis. MLA recommends one space but allows two spaces after a period, question mark, or exclamation point at the end of a sentence. Form dashes by using two hyphens with no spaces between them. Do not leave a space before or after a dash.

### Outline

MLA style does not require an outline. If your instructor requires an outline, however, inquire as to the style it should take. Present the outline on the page following the title page (if any).

### Page numbers

Place your last name and the page number (e.g., Smith 1) in the upper right corner of each page, approximately one-half inch from the top and one inch from the right edge of the page. Do not use the word *page* or its abbreviation *p.*; do not use a period or any other mark of punctuation with your name and page number. Number all pages of your paper, including the last.

### Works Cited page

The list of works cited is placed on a separate page at the end of your paper and titled Works Cited. Place your last name and page number in the upper right-hand corner, and center the words *Works Cited.* For a model Works Cited page from a student research paper, see page 202. For the specific requirements of the format of each entry in the list of works cited, refer to the model entries on pages 184–195.

To assemble a Works Cited list for your paper, follow these guidelines:

- Alphabetize your sources by the authors' (or editors') last names.
- If you have two or more authors with the same last name, alphabetize by the first name.
- If you have two or more works by the same author, alphabetize by the first word of the title, not counting *a,*

*an*, or *the*. Use the author's name in the first entry and three hyphens followed by a period in subsequent entries.

- If no author is known, alphabetize by title.
- Start the first line of each entry at the left margin, and indent subsequent lines one-half inch (or five spaces).
- Double-space both within and between entries.
- Separate the three main parts of an entry—author's name, title of work, and publication data—with periods followed by one space.
- Give the author's name, last name first. For sources with two or three authors, use normal order for all names after the first, using commas to separate the names.
- Provide complete titles, and capitalize all important words. Continuously underline titles of books and periodicals. Use quotation marks with titles of periodical articles, chapters and essays within books, short stories, and poems.
- With books, give place of publication, publisher, and date. You may shorten the publisher's name (Viking for Viking Press). Omit *Publisher, Co., Inc.*, etc., and use *UP* for *University Press.* (For sample MLA-style book entries, see pages 186–188.)
- With periodicals, give the volume or issue number, date (abbreviate months except for May, June, and July), and page numbers. (For sample MLA-style periodical entries, see pages 188–189.)

## SAMPLE RESEARCH PAPER PAGES

Recycling Used Newsprint:

Coming to Terms with an Environmental Problem

**Title is centered and placed about a third of the way down the page.**

by

Patricia LaRose

**Six lines below the title, center *by,* skip two lines, and center your name.**

English 001, Section G

Professor P. Eschholz

April 5, 2002

**Course information and date are centered and double-spaced.**

LaRose 1

Recycling Used Newsprint:

Coming to Terms with an Environmental Problem

**Title is centered.**

**Anecdotal introductory paragraph grabs readers' attention and involves them in the issue.**

Two decades ago, television viewers were shocked at the pictures of a barge load of trash from Babylon, New York, as it made its odyssey up and down the eastern seaboard. At every port it was turned away with the cry that there was "no room at the inn," while an entourage of seagulls nibbled away at its decaying cargo. It may have been the first time most Americans were forced to come face-to-face with the unpleasant fact that trash is one of our most pressing problems. The United States is a "throw-away society" according to Jim Hightower, commissioner of agriculture in Florida, that every year tosses out "41 million tons of food and yard waste, 13 million tons of metals, 12 million tons of glass, and 10 million tons of plastic" (xv-xvi).

**In-text citation with the author's name given in a signal phrase and the page number given in parentheses.**

For a long time Americans identified plastic and Styrofoam as the chief culprits, and perhaps justifiably so because they are not biodegradable. But with all the attention we have paid to them, we have ignored a problem that is even worse. Noel Grove reports that paper waste accounts for more than 70 million tons annually--approximately 37 percent of our nation's waste--and that 50 million tons wind up in landfills (112). This is enough paper waste to fill 2.5 million trailer trucks to the brim. The problem is not expected to lessen. The Environmental Protection Agency calculates that paper waste deposited in landfills will

**Paraphrase of a source with the author's name in a signal phrase and the page number in parentheses.**

LaRose 9

Works Cited

American Forest and Paper Association. "Old Newspapers." Recycling 21 May 1997. 9 Mar. 2001 <http://www.afandpa.org/Recycling/Paper/index.html#OldNewspapers>.

Blumberg, Louis, and Robert Gottlieb. War on Waste: Can America Win Its Battle with Garbage? Washington: Island, 1989.

Brown, Elizabeth A. "Paper Recycling Catches on Slowly." Christian Science Monitor 14 Nov. 1989: 12-13.

Carra, Joseph S. "Municipal Solid Waste and Sanitary Landfilling in the United States of America." International Perspectives on Municipal Solid Wastes and Sanitary Landfilling. Ed. Joseph S. Carra and Raffaello Cossu. San Diego: Academic, 1990. 221-34.

Easterbrook, Gregg. "Good News from Planet Earth." USA Weekend 14-16 Apr. 1995: 4-6.

- - -. A Moment on the Earth: The Coming Age of Environmental Optimism. New York: Viking, 1995.

Eller, Daryn. "Spare the Wrappings: Give Trees a Chance." Longevity Dec. 1990: 96.

Gore, Al. Earth in the Balance: Ecology and the Human Spirit. New York: Plume, 1993.

Grove, Noel. "Recycling." National Geographic July 1994: 92-115.

Hagerty, D. Joseph, Joseph L. Pavoni, and John E. Heer, Jr. Solid Waste Management. New York: Van Nostrand, 1973.

***Works Cited* is centered at the top of the page.**

**The list of works cited begins on a new page. The first line of each entry begins at the left margin; subsequent lines are indented five spaces. Double-space within entries as well as between entries.**

**The correct MLA forms for various other kinds of publications are given on pages 184–195.**

# 53 APA Documentation Style

In writing a research paper, you are using the information and ideas of others. Whenever you directly quote, summarize, or paraphrase another person's thoughts and ideas or use facts and statistics that are not commonly known or believed, you must document, or properly acknowledge, your source.

The American Psychological Association (APA) style uses brief in-text citations to refer the reader to full bibliographic information in a list of references at the end of the paper. This documentation style is used in psychology and other social sciences. The following recommendations are based on the *Publication Manual of the American Psychological Association,* fifth edition (2001), as well as on the APA's Web site <http://www.apa.org>.

## 53a APA in-text citations

An APA in-text citation gives the author's last name and the year of publication. In the case of a quotation, include the page number preceded by *p.* (a page number for a paraphrase or summary is optional). Use commas to separate items in the parentheses. When two or more citations are given in a single parenthesis, use a semicolon to separate the citations.

### Directory to APA In-Text Citations



(*continued*)

Directory (*continued*)



**1. PARAPHRASE OR SUMMARY WITH AUTHOR'S NAME IN TEXT**

Milstein (1990) believes that the grizzly bear, Dall sheep, and golden eagle are in danger if we do not curb poaching in our national parks.

**2. PARAPHRASE OR SUMMARY WITHOUT AUTHOR'S NAME IN TEXT**

The grizzly bear, Dall sheep, and golden eagle are in danger if we do not curb poaching in our national parks (Milstein, 1990).

**3. QUOTATION WITH AUTHOR'S NAME IN TEXT**

According to Milstein (1990), "as part of a recent Washington State wilderness bill, Congress upped the maximum fine for wildlife crimes at Olympic National Park from $500 to $25,000" (p. 125).

**4. QUOTATION WITHOUT AUTHOR'S NAME IN TEXT**

During one federal sting operation in North Carolina, "agents bought about 1,200 bear claws, on their way to becoming jewelry; 102 feet, an exotic delicacy; 320 gallbladders, considered a panacea in the Orient; 18 heads; 18 hides; and one live bear cub" (Milstein, 1990, p. 122).

**5. A WORK WITH TWO AUTHORS**

List both authors' names in text with the year of publication immediately following, or put both names in a parenthetical citation. In this case, note that the ampersand (&) is used instead of *and*.

The resurgence of recycling in the 1980s can be traced to the confrontations between incineration and landfill projects (Blumberg & Gottlieb, 1989).

**6. A WORK WITH THREE TO FIVE AUTHORS**

Give all authors' last names in the text or parentheses in your *first* citation only; use an ampersand (&) instead of *and* in the parenthetical citation.

Liberation begins with an awareness that we all are to some degree prisoners of our own language (Clark, Eschholz, & Rosa, 2000).

In subsequent citations, use only the first author's name and *et al.* (Latin for "and others").

Clark et al. (2000) maintain that language is one of humankind's greatest achievements and most important resources.

7. A WORK WITH SIX OR MORE AUTHORS

Cognitive behavior therapy is now recognized as an effective intervention for disturbed body image in young women (Rosen et al., 1989).

8. AUTHOR UNKNOWN

In *Children of the Dragon: The Story of Tiananmen Square* (1990), the interconnections of the events leading up to the bloody massacre in June 1989 are presented in detail.

The interconnections of the events leading up to the bloody massacre in China in June 1989 are presented in detail (*Children of the Dragon,* 1990).

9. CORPORATE AUTHOR

According to Internal Revenue Service (IRS) (1997) regulations, two new tax credits for education-related expenses exist.

Under the new regulations, there are two tax credits for education-related expenses (Internal Revenue Service [IRS], 1997).

In subsequent citations, the abbreviation is sufficient both in the text and parenthetically.

10. ONE OF THE TWO OR MORE WORKS BY THE SAME AUTHOR

When your reference list includes two or more works by the same author, the date of publication should be sufficient to distinguish them. If, however, you happen to have two or more works by the same author published in the same year, the works should be lettered (e.g., 1999a, 1999b, 1999c) in your reference list (see page 209) and cited that way in your parenthetical citation.

In the wild, young orangutans are inseparable from their mothers during their first six years of life (Townsend, 1999a).

11. TWO OR MORE AUTHORS WITH THE SAME LAST NAME

When using references by two or more authors with the same last name, include initials with all references.

> A survey completed by P. J. Babin (2001) showed that sales of cellular car phones tripled in the last five years.

12. MORE THAN ONE WORK IN A PARENTHETICAL CITATION

When acknowledging more than one source in a parenthetical citation, present the sources in the order in which they appear in your list of references (i.e., alphabetically); separate the citations with a semicolon.

> Environmentalists contend that solid waste disposal will be America's--if not the world's--most pressing problem in the future (Blumberg & Gottlieb, 1989; Rathje, 1989).

13. A SOURCE CITED IN ANOTHER SOURCE

It is always preferable to borrow material from original sources when they are available. If you need to cite a person quoted or paraphrased in a work written by another person, begin your parenthetical citation with the phrase *as cited in.*

> A study by F. Bryan indicates that women are increasingly participating in local and state government (as cited in Sherman, 1999).

Readers know from this citation that a reference to Bryan's work was made in Sherman.

14. AN ELECTRONIC SOURCE

Cite electronic sources as you would print sources, with the author's last name and date of posting. For direct quotations from electronic sources that number paragraphs and not pages, use the abbreviation *para* or the # symbol before the paragraph number. In the absence of any numbering, the author's name and the date will suffice.

> M. Gillespie (1999) points out that "opposition to the death penalty among minorities is higher than among whites" (para. 4).

15. PERSONAL COMMUNICATIONS

For telephone conversations, interviews, letters, memos, e-mail, and other unpublished personal communications, give the writer's initials and last name, the words *personal communication,* and the exact date.

> K. DeAngelus (personal communication, April 20, 2000) noted that the new city ordinances should not delay the planned expansion of the downtown Civic Center.

## 53b APA-style references

In APA style, the alphabetical list of works cited that appears at the end of a research paper is called *References.*

### Directory to APA List of References



(*continued*)

Directory *(continued)*



## Books

### 1. ONE AUTHOR

Katznelson, I. (1992). *Marxism and the city.* New York: Oxford University Press.

### 2. TWO OR MORE AUTHORS

Davison, J., & Davison, L. (1994). *To make a house a home: Four generations of American women and the houses they lived in.* New York: Random House.

Debold, E., Wilson, M., & Malave, I. (1993). *Mother daughter revolution: From good girls to great women.* New York: Bantam.

### 3. CORPORATE AUTHOR

National Association for the Advancement of Colored People. (1994). *Beyond the Rodney King story: An investigation of police conduct in minority communities.* Boston: Northeastern University Press.

### 4. AUTHOR UNKNOWN

*Children of the dragon: The story of Tiananmen Square.* (1990). New York: Macmillan.

5. **EDITOR**

Singer, R. N., Murphey, M., & Tennant, L. K. (Eds.). (1993). *Handbook of research on sport psychology.* New York: Macmillan.

6. **WORK IN AN ANTHOLOGY**

Pinderhughes, E. (1995). Biracial identity—Asset or handicap? In H. W. Harris, H. C. Blue, & E. E. H. Griffith (Eds.), *Racial and ethnic identity: Psychological development and creative expression* (pp. 163-179). New York: Routledge.

7. **EDITION OTHER THAN FIRST**

Polsby, N. W., & Wildavsky, A. (1991). *Presidential elections: Contemporary strategies of American electoral politics* (2nd ed.). New York: Free Press.

8. **TRANSLATION**

Kafka, F. (1984). *The penal colony: Stories and short pieces* (W. E. Muir, Trans.). New York: Schocken.

9. **MULTIVOLUME WORK**

Cassidy, F. G., & Hall, J. H. (Eds.). (1985). *Dictionary of American regional English* (Vols. 1-3). Cambridge, MA: Harvard University Press.

10. **TWO OR MORE WORKS BY THE SAME AUTHOR**

When listing two or more works by the same author, arrange them chronologically by year of publication, starting with the one published earliest. If two or more works by the same author were published in the same year, use lowercase letters to differentiate them: (1992a), (1992b), (1992c).

Sheehy, G. (1988). *Character: America's search for leadership.* New York: Morrow.

Sheehy, G. (1992). *The silent passage: Menopause.* New York: Random House.

Sheehy, G. (1995). *Passages 2000: The revolution of second adulthood.* New York: Random House.

## Periodicals

**11. ARTICLE IN A JOURNAL PAGINATED BY VOLUME**

Izraeli, D. N. (1993). They have eyes and see not: Gender politics in the Diaspora Museum. *Psychology of Women Quarterly, 17,* 515-523.

**12. ARTICLE IN A JOURNAL PAGINATED BY ISSUE**

Lachman, S. J. (1993). Psychology and riots. *Psychology: A Journal of Human Behavior, 30*(3/4), 16-23.

**13. ARTICLE IN A MONTHLY MAGAZINE**

Ross, J. F. (2000, December). People of the reindeer. *Smithsonian, 31,* 51–64.

**14. ARTICLE IN A WEEKLY MAGAZINE**

Peyser, M. (2001, January 1). Reality TV's real survivor. *Newsweek, 136,* 77.

**15. ARTICLE IN A NEWSPAPER**

Parlez, J. (2001, January 31). Mexico warns of Colombia drug war spillover. *The New York Times,* p. A8.

**16. EDITORIAL**

Cole, E., & Rothblum, E. D. (1991). [Editorial]. *Women and Therapy: A Feminist Quarterly, 11*(2), 1-2.

**17. LETTER TO THE EDITOR**

Seeley, D. S. (1991). "What's left?" is the wrong question [Letter to the editor]. *Social Policy, 21*(4), 60.

**18. BOOK REVIEW**

Retish, P. (1992). [Review of the book *The troubled adolescent*]. *Journal of Pediatric Psychology, 17,* 245-246.

## Electronic sources

Your goal with a reference to an electronic source is to both credit the author and enable the reader to retrieve the source. APA format stresses that you provide readers with as much specific information as you can, including the retrieval path you followed and a correct Internet address or URL. The following models are based on the lastest recommendations on the APA's Web site <http://www.apastyle.org/elecref.html>.

**19. ONLINE PUBLICATION WITH PRINT EQUIVALENT**

Schino, G. (2000). Grooming, competition, and social rank among female primates: A meta-analysis. [Electronic version]. *Animal Behavior, 62,* 265–271.

If you believe that the online version differs in any way from the print version, you should add the date you accessed the article and the URL.

Schino, G. (2000). Grooming, competition, and social rank among female primates: A meta-analysis. *Animal Behavior, 62,* 265–271. Retrieved October 7, 2001, from http://www.idealibrary.com/links/doi/10.1006/anbe.2001.1750/pdf

**20. ONLINE JOURNAL OR MAGAZINE WITHOUT PRINT EQUIVALENT**

Johnson, D. E. (2000, October). Helping children succeed after divorce: Building a community-based program in a rural county. *Journal of Extension, 38* (5). Retrieved January 30, 2001, from http://joe.org/joe/2000october/iwl.html

**21. ARTICLE RETRIEVED FROM DATABASE**

Brown, S. P., Ganesan, S., & Challagalla, G. (2001). Self-efficacy as a moderator of information-seeking effectiveness. *Journal of Applied Psychology, 86,* 1043–1051. Retrieved December 12, 2001, from PsychINFO database.

**22. ABSTRACT RETRIEVED FROM CD-ROM DATABASE**

Meyer, A. S., & Bock, K. (1992). The tip-of-the-tongue phenomena: Blocking or partial activation? [CD-ROM]. *Memory and Cognition, 20,* 715–726. Abstract retrieved January 11, 2002, from PsycLIT database.

**23. ONLINE NEWSPAPER ARTICLE**

Elliot, S. (1996, July 22). Nielsen unit offers data about internet users. *The New York Times.* Retrieved February 2, 2001, from http://www.nytimes.com/library/articles/users.html

24. WORLD WIDE WEB SITE

Library of Congress. (1998). Jackie Robinson and other baseball highlights 1860s–1960s. *American Memory Project.* Retrieved January 7, 2002, from http://memory.loc.gov/ammem/jrhtml/jrhome.html

25. ELECTRONIC DISCUSSION LIST MESSAGE

Preston, D. R. (1997, November 8). Basketball terms. Message posted American Dialect Society electronic mailing list, archived at http://www2.et.byu.edu/~lilliek/ads/indes.htm

## Other sources

26. GOVERNMENT PUBLICATION

U.S. Department of Defense. (1997). *Proliferation: Threat and response* (DD Publication No. ADM 90-1679). Washington, DC: U.S. Government Printing Office.

27. DISSERTATION ABSTRACT

Erwin, L. K. (1990). The politics of anti-feminism: The pro-family movement in Canada (Doctoral dissertation, York University, 1990). *Dissertation Abstracts International, 51,* 3237-A.

28. PUBLISHED CONFERENCE PROCEEDINGS

Hanks, T. (Ed.). (1999). *The great American writers: Proceedings of the eleventh annual research symposium on literature, Boston, 21–23 April 1999.* Boston: University of Massachusetts Press.

29. PUBLISHED INTERVIEW

Chatzky, J. S. (2000, May–June). Talking money [Interview with Katie Couric]. *Money for Women,* pp. 60–62.

Unpublished interviews like the ones you may conduct while doing your research are not included in your list of references. They are, however, cited parenthetically in the text of your paper as illustrated in the model for personal communications on page 206–207.

30. COMPUTER SOFTWARE

Q-Corp. (1997). *Q-Notes: Electronic note-taking* [Computer software]. Needham Heights, MA: Allyn & Bacon.

31. VIDEOTAPE OR RECORDING

Hutchison, S., & Carpenter, K. (Producers), & Wheeler, C. (Director). (1993). *Nez Perce: I will fight no more forever* [Videotape]. Alexandria, VA: Time-Life Video.

Flanders, D. (1997). *Mother make my bed: Vermont folk songs and ballads from the Flanders collection* [CD]. Burlington, VT: The Vermont Performing Arts League.

## 53c APA manuscript format

The American Psychological Association recommends using the following guidelines for formatting manuscripts.

### Paper and type

For academic papers use 8 1/2" × 11", twenty-pound white paper, and type or print in black on one side of each sheet. Most instructors ask you to use a standard type style such as Times Roman or Courier, preferring that you avoid italic or script type styles for your main text. A quality printer such as an ink-jet or laser gives the sharpest printout. If you use a dot matrix printer, be sure to use its letter-quality setting and a fresh ribbon. If you use continuous-feed paper, remove the perforated strips from the sides of the paper, separate the pages, and use a paper clip (do not staple) to secure the pages. Finally, be sure you keep both a paper copy and an electronic copy of your paper.

### Margins, spacing, and indentation

Leave a one-inch margin on all sides of the page. Some instructors prefer a one-and-one-half-inch margin on the left. Double-space the text of the paper, including set-off quotations and entries on the References page. Do not justify (make even) the right-hand margin. Indent the first line of each paragraph five spaces.

APA style requires that quotations of more than forty words be indented five to seven spaces from the left margin and double-spaced. Do not use quotation marks.

### Page numbers and short title

Use arabic numerals (1, 2, 3 . . .) for page numbers together with a shortened title (usually the first three words) on each manuscript page. The header and the page number should be separated by five spaces. The page number should be one inch from the right edge of the paper. Do not use the word *page* or its abbreviation *p.*; do not use a period or any other mark of punctuation. Number all pages of your paper, including the title page and the References page.

### Title page

APA formatting requires a numbered title page. The page number should be placed one-half inch from the top of the page and one inch from the right edge of the page. Before the page number, and separated from it by five spaces, type a shortened title (usually the first three words of the title). Center the title on the page about fifteen lines from the top. Double-space, and center your name and the course information and the date directly under the title. See model title page on page 217.

Do not underline or italicize, put quotation marks around, or put a period at the end of the title of your paper. Capitalize the first letter of all words in your title except articles, coordinating conjunctions, and prepositions, unless they are the first or last word in a title or contain four letters or more.

### Outline or abstract

APA format does not require an outline; however, some instructors who follow APA format do require an abstract, or brief summary (not exceeding 120 words), of your paper. Your abstract should be a statement of your thesis, the major points of your argument, and the conclusions you have drawn. The abstract should be placed after the title page and numbered as page 2. The word *Abstract* should appear centered at the top margin.

### Spacing for punctuation

APA requires that you leave one space after all punctuation and between the three periods in an ellipsis. Form dashes by using two hyphens. Do not leave a space before or after a dash.

### Headings and subheadings

APA has specific requirements for headings and subheadings. For major headings, use upper- and lowercase, and center the heading. For subheadings, start at the left-hand margin and italicize or underline; use upper- and lowercase. Do not number the headings. Capitalize all major words in both heading and subheadings.

### References page

The alphabetical list of works that you cite in your research paper is placed on a separate page at the end of your paper and titled References. Place the short title of your paper and the page number in the upper right-hand corner of the page. Double-space, and center the word *References*. Double-space within entries and between entries. For a model References page from a student research paper, see page 219. For specific format requirements for the various types of entries in your references, refer to the model entries on pages 207–213.

To assemble a list of sources for an APA-style paper, follow these guidelines:

- Alphabetize your sources by the authors' (or editors') last names.
- If you have two or more authors with the same last name, alphabetize by the first initials.
- If you have two or more works by the same author, arrange them chronologically by year of publication, starting with the one published earliest. If two or more of these works were published in the same year, use lowercase letters to differentiate them (1998a), (1998b), and (1998c). Repeat the author's name including any initials for all entries.
- If no author is known, alphabetize by the first word of the title not counting *a*, *an*, or *the*.
- Reverse *all* author's names within each entry, and use initials, not first names.
- Name all authors; do not use *et al.* Use an ampersand (&) instead of *and* in naming the second or last of several authors.
- Give the date of publication in parentheses after the last author's name, followed by a period.
- Italicize the titles and subtitles of the books and periodicals, but do not put quotation marks around titles of articles.

- Capitalize only the first word of book and article titles and subtitles and any proper nouns. Capitalize the titles of periodicals according to standard rules.
- When referring to a state in the place of publication, use postal service abbreviations (e.g., Springfield, IL:).
- Provide the full name of university presses, corporations, and professional associations serving as publishers. Abbreviate the names of commercial publishers, omitting such terms as *Publishers*, *Co.*, and *Inc.*; retain the term *Books* and, for academic publishers, *Press*.
- Use the abbreviations *p.* or *pp.* for page numbers of all newspaper articles and chapters in books.
- Individual entries should be presented with a hanging indent. Start the first line of each entry at the left margin, indenting subsequent lines five spaces.

## SAMPLE RESEARCH PAPER PAGES

Public Opinion 1

**Identification heading: short title and page**

Public Opinion and the Death Penalty

Brooke Bailey

Criminal Justice 4700: Introduction to Research Methods

Professor S. Eschholz

July 27, 2002

**Full title of paper, writer's name, course information, and date are centered and double-spaced.**

Short title and page number

Public Opinion 2

Title is centered.

Public Opinion and the Death Penalty

Authors' names given in text followed by date in parentheses.

For a direct quotation, the writer gives the page number preceded by *p.*

Questions highlight main issues the paper will address.

When author's name is not given in the text, it is given parenthetically with the date.

Public opinion greatly influences decisions made by the Supreme Court. In fact, public opinion even influences which cases the Court will address. Is this good or bad? On the one hand,"we the people" are the backbone of our court system. Murphy and Tanehaus (1990) believe that "the stability, even the continual existence, of a constitutional democracy depends heavily on public opinion" (p. 991). On the other hand, public opinion changes with the times--with changes in social and economic conditions. The question, then, is, to what extent should the Supreme Court be guided by public opinion? And specifically, to what extent should Supreme Court rulings on the death penalty--a very subjective issue--be guided by public opinion?

Let's look first at the question of public opinion and the death penalty. Gallup poll statistics from 1936 to 1994 show a stable and currently increasing level of support for the death penalty. The only exception was a brief period in the early 1960s, likely due to the new civil rights movement (Erskine, 1970). In the 1960s, a proportionally small but very vocal segment of the population had a lot to say about civil rights issues. During this period, the Supreme Court heard an abundance of civil rights cases, which introduced the idea that a vocal minority can have just as much influence when it comes to public opinion as a surveyed majority.

Public Opinion 8

References

Branham, L.S. (1994). *Sentencing, corrections, and prisoners' rights*. St. Paul, MN: West.

Casper, J.D. (1976). The Supreme Court and national policy making. *American Political Science Review, 70* (1), 50-63.

Erskine, H. (1970). The polls: Capital punishment. *Public Opinion Quarterly*, *34*, 290.

Marshall, T. (1987). The Supreme Court as an opinion leader: Court decisions and the mass public. *American Politics Quarterly*, *15*, 147.

Murphy, W., & Tanehaus, J. (1990). Publicity, public opinion, and the Court. *Northwestern University Law Review*, *84*, 995-1020.

Sarat, A., & Vidmar, N. (1976). Public opinion, the death penalty, and the Eighth Amendment: Testing the Marshall hypothesis. *Wisconsin Law Review*, *40*, 171.

Schmalleger, F. (1993). *Criminal justice today*. Upper Saddle River, NJ: Prentice Hall, 1993.

Vidmar, N., & Ellsworth, P. (1974). Public opinion and the death penalty. *Stanford Law Review*, *26*, 1245-1270.

Wrightsman, L.S., Nietzel, M., & Fortune, W. H. (1994). *Psychology and the legal system*. Pacific Grove, CA: Brooks/Cole.

**The heading *References* is centered at the top of the page.**

**Writer follows APA guidelines for list of references.**

**The list begins on a new page, and all entries are given in alphabetical order by authors' last names.**

**The first line of each entry begins at the left margin; subsequent lines are indented five spaces.**

**Writer double-spaces within entries as well as between entries.**

**The correct APA forms for various other kinds of publications are given on pages 207–213.**

# 54 Chicago Manual Documentation Style

Widely used in history, art history, philosophy, and some other disciplines in the humanities, *The Chicago Manual of Style* and *A Manual for Writers of Term Papers, Theses, and Dissertations* recommend the use of endnotes or footnotes rather than in-text citations. Present endnotes together at the "end" of the paper; present footnotes on the bottom, or "foot," of the page on which the citations occur. Use a raised Arabic numeral immediately after the item being acknowledged. Number endnotes or footnotes consecutively throughout the paper.

The first time you cite a source, provide complete publication data including the page number for the material you are quoting, paraphrasing, or summarizing. For subsequent references, you can use a shortened version of the citation. Indent the first line of each note five spaces, and bring all subsequent lines to the left margin. Single-space within each note, and double-space between notes.

## 54a *Chicago Manual* endnotes (or footnotes) and bibliography entries

The following are model citations for the most commonly used research sources. The endnote and bibliography entry for each source appear together for easy access.

### Directory to *Chicago Manual* Note and Bibliography Models

**Books**

## Books

### 1. ONE AUTHOR

1. Michael Ferber, The Poetry of Shelley (New York: Penguin Books, 1993), 173.

Ferber, Michael. The Poetry of Shelley. New York: Penguin Books, 1993.

### 2. TWO OR THREE AUTHORS

2. Kate Fullbrook and Edward Fullbrook, Simone de Beauvoir and Jean-Paul Sartre: The Remaking of a Twentieth Century Legend (New York: Basic Books, 1994), 119.

Fullbrook, Kate, and Edward Fullbrook. Simone de Beauvoir and Jean-Paul Sartre: The Remaking of a Twentieth Century Legend. New York: Basic Books, 1994.

3. FOUR OR MORE AUTHORS

3. Harold Bloom et al., Deconstruction and Criticism (New York: Seabury Press, 1979), 133.

Bloom, Harold, Paul de Man, Geoffrey H. Hartman, and J. Hillis Miller. Deconstruction and Criticism. New York: Seabury Press, 1979.

4. AUTHOR WITH AN EDITOR

4. William Shakespeare, Antony and Cleopatra, ed. Michael Neill (New York: Oxford University Press, 1994), 71.

Shakespeare, William. Antony and Cleopatra. Edited by Michael Neill. New York: Oxford University Press, 1994.

5. EDITOR

5. A. Wayne Glowka and Donald M. Lance, eds., Language Variation in North American English: Research and Teaching (New York: Modern Language Association of America, 1993), 354.

Glowka, A. Wayne, and Donald M. Lance, eds. Language Variation in North American English: Research and Teaching. New York: Modern Language Association of America, 1993.

6. AUTHOR WITH A TRANSLATOR

6. Franz Kafka, The Penal Colony: Stories and Short Pieces, trans. Willa Muir and Edwin Muir (New York: Schocken Books, 1984).

Kafka, Franz. The Penal Colony: Stories and Short Pieces. Translated by Willa Muir and Edwin Muir. New York: Schocken Books, 1984.

7. UNKNOWN AUTHOR

7. The Vermont Almanac (Middlebury, Vt.: Regional Facts, 1993), 17–26.

The Vermont Almanac. Middlebury, Vt.: Regional Facts, 1993.

8. EDITION OTHER THAN FIRST

8. William Zinsser, On Writing Well, 5th ed. (New York: HarperPerennial, 1994), 70–72.

Zinsser, William. On Writing Well. 5th ed. New York: HarperPerennial, 1994.

9. WORK IN MORE THAN ONE VOLUME

9. J. B. Schneewind, Moral Philosophy from Montaigne to Kant: An Anthology (New York: Cambridge University Press, 1990), 1:630.

Schneewind, J. B. Moral Philosophy from Montaigne to Kant: An Anthology. Vol. 1. New York: Cambridge University Press, 1990.

10. TITLED VOLUME IN A MULTIVOLUME WORK

10. Hannah Arendt, Imperialism, vol. 2 of The Origins of Totalitarianism (London: Harcourt Brace Jovanovich, 1968), 153–55.

Arendt, Hannah. Imperialism. Vol. 2 of The Origins of Totalitarianism. London: Harcourt Brace Jovanovich, 1968.

#### 11. WORK FROM AN ANTHOLOGY

11. Rita Dove, "Parsley," in New Worlds of Literature, ed. Jerome Beaty and J. Paul Hunter (New York: W. W. Norton, 1989), 502.

Dove, Rita. "Parsley." In New Worlds of Literature, edited by Jerome Beaty and J. Paul Hunter, 502. New York: W. W. Norton, 1989.

#### 12. WORK IN A SERIES

12. Frances M. Frost, Hemlock Wall, The Yale Series of Younger Poets, no. 27 (New Haven: Yale University Press, 1929), 11.

Frost, Frances M. Hemlock Wall. The Yale Series of Younger Poets, no. 27. New Haven: Yale University Press, 1929.

#### 13. ARTICLE IN A REFERENCE WORK

The abbreviation *s.v.* stands for the Latin phrase *sub verbo*, literally, "under the word."

13. The American Heritage Dictionary, 4th ed., s.v. "culture."

Reference works such as dictionaries and encyclopedias are usually not listed in the bibliography.

13. Francesco Mazzoni, "Dante," in The New Encyclopaedia Britannica: Macropaedia, 16th ed.

## Periodicals

#### 14. ARTICLE IN A JOURNAL PAGINATED BY VOLUME

14. Harold Fromm, "My Science Wars," Hudson Review 49 (1997): 601.

Fromm, Harold. "My Science Wars." Hudson Review 49 (1997): 599-609.

#### 15. ARTICLE IN A JOURNAL PAGINATED BY ISSUE

15. Michael Lobel, "Warhol's Closet," Art Journal 55, no. 4 (1996): 47.

Lobel, Michael. "Warhol's Closet." Art Journal 55, no. 4 (1996): 42-50.

#### 16. ARTICLE IN A MAGAZINE

16. David Zane Mairowitz, "Fascism à la Mode," Harper's, October 1997, 61.

Mairowitz, David Zane. "Fascism à la Mode." Harper's, October 1997, 59-67.

#### 17. UNSIGNED ARTICLE

17. "Pompeii: Will the City Go from Dust to Dust?" Newsweek, 1 September 1997, 8.

"Pompeii: Will the City Go from Dust to Dust?" Newsweek. 1 September 1997, 8.

**18. ARTICLE IN A NEWSPAPER**

18. William J. Broad, "Another Possible Climate Culprit: The Sun," New York Times, 23 September 1997, C1.

Broad, William J. "Another Possible Climate Culprit: The Sun." New York Times, 23 September 1997, C1.

**19. A REVIEW**

19. Christopher Camuto, "Amphibians, Aliens, and Agriculture," review of Nature Out of Place: Biological Invasions in the Global Age, by Jason Van Driesche and Roy Van Driesche, Audubon, January-February 2001, 115.

Camuto, Christopher. "Amphibians, Aliens, and Agriculture." Review of Nature Out of Place: Biological Invasions in the Global Age, by Jason Van Driesche and Roy Van Driesche. Audubon, January-February 2001, 115.

19. Morris Dickstein, "Sounds of Silence," review of Rethinking the Holocaust, by Yehuda Bauer. New York Times Book Review, 28 January 2001, 10.

Dickstein, Morris. "Sounds of Silence." Review of Rethinking the Holocaust, by Yehuda Bauer. New York Times Book Review, 28 January 2001, 10.

## Electronic sources

*The Chicago Manual of Style* (14th ed., 1993) provides little guidance for documenting electronic sources, but Kate L. Turabian's *A Manual for Writers of Term Papers, Theses, and Dissertations* (6th ed., 1996) offers some basic models for handling electronic sources in *Chicago* style, and these models are presented below. *The Chicago Manual of Style* is not scheduled to be updated until 2003. In the meantime, for other electronic sources and more up-to-date, easier-to-use formats, your instructor may wish you to use the model citations provided by *Columbia Online Style* at its Web site <http://www.columbia.edu/cu/cup/cgos/idx_basic.html>, or the system developed by Andrew Harnack and Eugene Kleppinger in *Online! A Reference Guide to Using Internet Sources* (1998).

**20. A SOURCE ON A PERIODICAL CD-ROM**

20. Caryn James, "An Army Family as Strong as Its Weakest Link," in New York Times, 16 September 1994, C8, New York Times Ondisc [CD-ROM], UMI-ProQuest, October 2000.

James, Caryn. "An Army Family as Strong as Its Weakest Link." New York Times, 16 September 1994, C8. New York Times Ondisc [CD-ROM], UMI-ProQuest, October 2000.

#### 21. A SOURCE ON A NONPERIODICAL CD-ROM

21. Donald Sheehy, ed. Robert Frost: Poems, Life, Legacy [CD-ROM] (New York: Henry Holt, 1997).

Sheehy, Donald, ed., Robert Frost: Poems, Life, Legacy [CD-ROM]. New York: Henry Holt, 1997.

#### 22. ONLINE BOOK

22. Nathaniel Hawthorne, Twice-Told Tales [book online], ed. George Parsons Lathrop (Boston: Houghton, 1883, accessed 1 March 2001); available from http://www.tiac.net/users/eldred/nh/ttt.html; Internet.

Hawthorne, Nathaniel. Twice-Told Tales [book online]. Edited by George Parsons Lathrop. Boston: Houghton, 1883. Accessed 1 March 2001. Available from http://www.tiac.net/users/eldred/nh/ttt.html; Internet.

#### 23. ARTICLE IN AN ONLINE PERIODICAL

23. Brian Doan. "Death Penalty Policy, Statistics, and Public Opinion," Focus on Law Studies 12, no. 2 (1997): par. 5, wysiwyg://44//http://www.abanet.org/publiced/focus/spr97pol.html (11 February 2001).

Doan, Brian. "Death Penalty Policy, Statistics, and Public Opinion." Focus on Law Studies 12, no. 2 (1997): 7 pars. wysiwyg://44//http://www.abanet.org/publiced/focus/spr97pol.html (11 February 2001).

## Other sources

#### 24. GOVERNMENT PUBLICATION

24. U.S. Bureau of the Census, Poverty in the United States, Current Population Reports, Ser. P-60, No. 163 (Washington, D.C.: GPO, 1989), 33.

U.S. Bureau of the Census. Poverty in the United States, Current Population Reports, Ser. P-60, No. 163. Washington, D.C.: GPO, 1989.

#### 25. PUBLISHED LETTER

25. Dorothy Canfield Fisher to Arthur L. Guerard, 9 September 1931, Keeping Fires Night and Day: Selected Letters of Dorothy Canfield Fisher, ed. Mark J. Madigan (Columbia: University of Missouri Press, 1993), 317.

Fisher, Dorothy Canfield. Letter to Arthur L. Guerard, 9 September 1931. In Keeping Fires Night and Day: Selected Letters of Dorothy Canfield Fisher, edited by Mark J. Madigan, 317. Columbia: University of Missouri Press, 1993.

**26. PUBLISHED INTERVIEW**

26. Paul Ridker, "New Clue to an Old Killer: Inflammation and Heart Disease," interview by Bonnie Liebman, Nutrition Action Newsletter 27, no. 7 (2000): 4.

Ridker, Paul. "New Clue to an Old Killer: Inflammation and Heart Disease." Interview by Bonnie Liebman. Nutrition Action Newsletter 27, no. 7 (2000): 3–5.

**27. PERSONAL LETTER OR INTERVIEW**

27. Christopher Bohjalian, letter to author, 4 September 2000.

Bohjalian, Christopher. Letter to author. 4 September 2000.

**28. WORK OF ART**

28. Sandro Botticelli, Birth of Venus, oil painting. Uffizi Gallery, Florence.

Botticelli, Sandro. Birth of Venus, oil painting. Uffizi Gallery, Florence.

**29. FILM OR VIDEOTAPE**

29. Forrest Gump, dir. Robert Zemeckis, with Tom Hanks and Sally Field, 2 hr. 15 min., Paramount, 1994, videocassette.

Forrest Gump, Directed by Robert Zemeckis, with Tom Hanks and Sally Field, 2 hr. 15 min., Paramount, 1994, videocassette.

**30. SOUND RECORDING**

30. Joseph Haydn, Concerto for Piano and Orchestra in F Major, Franz Liszt Chamber Orchestra, Emanuel Ax, Sony SK 48 383, 1992, compact disk.

Haydn, Joseph. Concerto for Piano and Orchestra in F Major. Franz Liszt Chamber Orchestra. Emanuel Ax. Sony SK 48 383, 1992. Compact disk.

## 54b Subsequent references

After you cite a source for the first time in a paper and fully document it, all subsequent references to that source are shortened. For most notes, give the author's last name followed by a comma and the page number. If no author is given, use a shortened title.

31. Dove, 502.

32. Vermont Almanac, 225.

If you cite two or more works by the same author or works by authors with identical last names, include a shortened title.

33. Emerson, "Literature and Change," 10.

34. Emerson, Representative Men, 17.

Use the abbreviation *Ibid.* (Latin for "in the same place") to refer to the work cited in the previous note. When citing the same page, you should use *Ibid.* alone. When citing a different page you must include the page number.

35. Ibid.

36. Ibid., 217.

## 54c *Chicago Manual* research paper format

*The Chicago Manual of Style,* fourteenth edition, and *A Manual for Writers of Term Papers, Theses, and Dissertations*, sixth edition, recommend using the following guidelines for formatting research papers.

### Paper and type

For academic papers use 8 1/2" × 11", twenty-pound white paper, and type or print in black on one side of each sheet. Most instructors ask you to use a standard type style such as Times Roman or Courier, preferring that you avoid italic or script type styles as your main text. A letter-quality printer, such as a laser printer, gives the cleanest printout. If you use a dot matrix printer, be sure to use its letter-quality setting and a fresh ribbon. If you use continuous-feed paper, remove the perforated strips from the sides of the paper, separate the pages, and use a paper clip (do not staple) to secure the pages. Finally, be sure you keep both a paper copy and an electronic copy of your paper.

### Title page, title, name, and course information

On the title page, place the title just above the center of the page, and center it. Four lines below the title, center the word *By* and your name two lines below it. The course title, professor's name, and date are all centered and double-spaced so as to allow a margin of one-and-one-half inches at the bottom of the page. Although you do not number the title page, count it and start numbering with page 2. See the sample on page 230.

Do not underline, put quotation marks around, or put a period at the end of the title of your paper. Capitalize all words in the title.

### Margins, spacing, and indentation

Leave a one-inch margin on all sides of the page. Some instructors prefer a one-and-one-half-inch margin on the left. Double-space the text of the paper, including set-off quotations. Single-space within, but double-space between, the entries on the Notes and Bibliography pages. Do not justify (make even) the right-hand margin. Indent the first line of each paragraph five spaces.

Set off quotations of more than ten typed lines of prose or two or more lines of poetry. Indent ten spaces from the left margin. Double-space before, after, and within the quotation. Do not use quotation marks.

### Spacing for punctuation

*Chicago Manual* style requires that you leave one space after all punctuation and between the three periods in an ellipsis. Form dashes by using two hyphens. Do not leave a space before or after a dash.

### Page numbers

Place the page number in the upper right corner of each page, approximately one-half inch from the top and one inch from the right edge of the page. Do not use the word *page* or its abbreviation *p.*; do not use a period or any other mark of punctuation with the page number. Number all pages of your paper except the title page; the first page of your text should be page 2.

### Notes page

The endnotes for a *Chicago*-style research paper are placed on a separate page at the end of your paper and titled Notes. Place your last name and page number in the upper right-hand corner, and center the word *Notes* in all-capital letters about an inch from the top of the page. Indent the first line of each entry five spaces, type the arabic numeral corresponding to the number in the text, and follow the numeral with a period and a space. Bring all subsequent lines flush left to the margin. Single-space within entries, and double-space between entries. For a model *Chicago*-style Notes page from a student research paper, see page 232. For the specific requirements of the format of each entry in the Notes, refer to the model entries on pages 221–226.

### Bibliography page

For a model *Chicago*-style bibliography page from a student research paper, see page 232. For the specific requirements of the format of each entry in the bibliography, refer to the model entries on pages 221–226.

To assemble a *Chicago*-style bibliography, follow these guidelines:

- Alphabetize your sources by the authors' (or editors') last names.
- If you have two or more authors with the same last name, alphabetize by the first name.
- If you have two or more works by the same author, alphabetize by the first word of the title, not counting *a, an,* or *the.* Use the author's name in the first entry and three hyphens followed by a period in subsequent entries.
- If no author is known, alphabetize by title.
- Start the first line of each entry at the left margin, and indent subsequent lines five spaces.
- Separate the three main parts of an entry—author's name, title of work, and publication data—with periods followed by one space.
- Give the author's name, last name first. For sources with two or three authors, use normal order for all names after the first, using commas to separate the names with an *and* before the last author's name.
- Provide complete titles, and capitalize all important words. Italicize or continuously underline titles of books and periodicals. Use quotation marks with titles of periodical articles, chapters and essays within books, short stories, and poems.

## SAMPLE RESEARCH PAPER PAGES

**No page number appears on title page.**

**Title is centered and capitalized, followed by the word *By* and the writer's name, both centered.**

THE ENGLISH-ONLY MOVEMENT:

CAN AMERICA PROSCRIBE LANGUAGE

WITH A CLEAR CONSCIENCE?

By

Jake Jamieson

**Course title, instructor's name, and date are centered and double-spaced.**

English 104

Professor Rosa

20 April 2002

2 **Page number**

A common conception among many people in this country is that the United States is a giant cultural "melting pot." For these people, the melting pot is a place where people from other places come together and bathe in the warm waters of assimilation. For many others, however, the melting pot analogy doesn't work. They see the melting pot as a giant cauldron into which immigrants are placed; here their native cultures, values, and backgrounds are boiled away in the scalding waters of discrimination.[1] One major point of contention in this debate is whether immigrants should be pushed toward learning English or encouraged to retain their native tongues.

**Raised arabic numerals for endnotes**

Those who argue that the melting pot analogy is valid believe that people who come to the United States do so willingly and should be expected to become a part of its culture instead of hanging on to their past. For them, the expectation that people who come to this country celebrate this country's holidays, dress as we do, embrace our values, and most important, speak our language is not unreasonable. They believe that assimilation offers the only way for everyone in this country to live together in harmony and the only way to dissipate the tensions that inevitably arise when cultures clash.[2] A major problem with this argument, however, is that no one seems to be able to agree on what exactly constitutes "our way" of doing things.

**Indent five spaces for new paragraph.**

Not everyone in America is of the same religious persuasion or has the exact same set of values, and

Page number

The heading NOTES is centered 1" from top of page.

Endnotes begin on a new page. The first line of each note is indented five spaces.

Writer single-spaces within notes and double-spaces between notes.

9

NOTES

1. David Price, "English-Only Rules: EEOC Has Gone Too Far," USA Today, 28 March 1996, A13.

2. William F. Buckley, "Se Habla Ingles: English as the Official American Language," National Review, 9 October 1995, 70.

3. Buckley, 71.

4. Craig Donegan, "Debate over Bilingualism: Should English Be the Nation's Official Language?" CQ Researcher, 19 January 1996, 52.

5. Mauro Mujica and Robert Underwood, "Should English Be the Official Language of the United States?" CQ Researcher, 19 January 1996, 65.

6. Donegan, 57.

7. Price, A13.

8. "English-Only Law Faces Test," Burlington Free Press, 26 March 1996, A1.

Page number

The heading BIBLIOGRAPHY is capitalized and centered.

The list begins on a new page, and all entries are given alphabetically.

Writer single-spaces within notes and double-spaces between notes.

10

BIBLIOGRAPHY

Buckley, William F. "Se Habla Ingles: English as the Official American Language." National Review, 9 October 1995, 70-71.

Donegan, Craig. "Debate over Bilingualism: Should English Be the Nation's Official Language?" CQ Researcher, 19 January 1996, 51-71.

"English-Only Law Faces Test," Burlington Free Press, 26 March 1996, A1.

Mujica, Mauro, and Robert Underwood, "Should English Be the Official Language of the United States?" CQ Researcher, 19 January 1996, 65.

Price, David, "English Only Rules: EEOC Has Gone Too Far." USA Today, 28 March 1996, A13.

# 55 CSE Documentation Style

Widely used in the life sciences, physical sciences, and mathematics, *Scientific Style and Format: The CBE Style Manual for Authors, Editors, and Publishers* (6th ed., 1994) and the CSE Web site <www.councilscienceeditors.org> recommend two systems of in-text citation: the name-year system and the citation-sequence system. In-text citations from both systems refer to a list of references at the paper's conclusion. Your instructor will tell you which system to use in your paper.

## 55a CSE name-year system

With a CSE name-year in-text citation, provide the author's last name and the year of publication in parentheses.

> Cutting saturated fat and cholesterol makes the biggest dent in LDL (Liebman 2001).

If the author's name is mentioned in the text, list only the publication year parenthetically.

> According to Liebman (2001), cutting saturated fat and cholesterol makes the biggest dent in LDL.

When a source has two authors, use "and" between the names: (Desmond and Moore 1991). When there are three or more authors, use "and others" with the first author's name: (Clark and others 1999). And when there is no known author, use Anonymous: (Anonymous 2000).

Provide complete information for each source in the alphabetically arranged list of references at the end of your paper.

## 55b CSE citation-sequence or numbered system

With the CSE sequence or numbered system, you assign numbers to your sources in the order in which they are first cited in your text. For example,

> Lutein, a carotenoid that colors spinach, kale, and collards, may help keep arteries from clogging.[4]

A subsequent citation of a previously cited source carries the same number as the initial citation. For multiple citations, give the sources' numbers, separated by commas:

Three recent studies[2,7,11] support this conclusion.

## 55c CSE reference list

On a separate page at the end of your paper titled References, or Cited References, give a list of the references you actually cited. Single-space within each entry, and double-space between entries on your list. If you selected the name-year system for your in-text citations, list your references alphabetically by authors' last names. With the citation-sequence system, list your references in the numerical order in which you cited them. With both systems start the first line of each entry at the left margin and indent subsequent lines 5 spaces. Use the following model entries to prepare your References page. For each type of source, we provide illustrations of both a name-year and a number reference.

### Books

**1. BOOK WITH ONE AUTHOR**

Pool R. 2001. Fat: fighting the obesity epidemic. New York: Oxford University Pr. 292 p.

1. Pool R. Fat: fighting the obesity epidemic. New York: Oxford University Pr; 2001. 292 p.

**2. BOOK WITH TWO OR MORE AUTHORS**

Desmond A, Moore J. 1991. Darwin: the life of a tormented evolutionist. New York: WW Norton. 808 p.

2. Desmond A, Moore J. Darwin: the life of a tormented evolutionist. New York: WW Norton; 1991. 808 p.

**3. BOOK WITH A CORPORATE AUTHOR**

Rand McNally. 2001. Atlas of the world. Millennium ed. New York: MetroBooks. 208 p.

3. Rand McNally. Atlas of the world. Millennium ed. New York: MetroBooks; 2001. 208 p.

#### 4. BOOK WITH AN EDITOR

Casey N. editor. 2001. Unholy ghost: writers on depression. New York: William Morrow. 288 p.

4. Casey N, editor. Unholy ghost: writers on depression. New York: William Morrow; 2001. 288 p.

#### 5. SECTION OF A BOOK WITH EDITOR(S)

Parens E. 2001. On the ethics and politics of embryonic stem cell research. In: Holland S, Lebucqz K, Zoloth L, editors. The human embryonic stem cell debate: science, ethics, and public policy. Cambridge: MIT Pr. p 37–50.

5. Parens E. On the ethics and politics of embryonic stem cell research. In: Holland S, Lebucqz K, Zoloth L, editors. The human embryonic stem cell debate: science, ethics, and public policy. Cambridge: MIT Pr; 2001. p 37–50.

#### 6. CHAPTER OF A BOOK

Piel G. 2001. The age of science: what scientists learned in the 20th century. New York: Basic Books. Chapter 7, Tools and human evolution; p 377–443.

6. Piel G. The age of science: what scientists learned in the 20th century. New York: Basic Books; 2001. Chapter 7, Tools and human evolution; p 377–443.

#### 7. PUBLISHED PROCEEDINGS OF A CONFERENCE

Scott DJ, editor. 2000. Climate change communication: proceedings of an international conference; 2000 June 22–24; Ottawa: Environment Canada. 324 p.

7. Scott DJ, editor. Climate change communication: proceedings of an international conference; 2000 June 22–24; Ottawa: Environment Canada; 2000. 324 p.

### Periodicals

#### 8. ARTICLE IN A JOURNAL PAGINATED BY VOLUME

Glasser FP. 2001. Mineralogical aspects of cement in radioactive waste disposal. Mineralog Mag 65:621–33.

8. Glasser FP. Mineralogical aspects of cement in radioactive waste disposal. Mineralog Mag 2001; 65:621–33.

#### 9. ARTICLE IN A JOURNAL PAGINATED BY ISSUE

Braatz SM. 1997. State of the world's forests 1997. Nature & Resources 33(3–4):18–25.

9. Braatz SM. State of the world's forests 1997. Nature & Resources 1997; 33(3–4):18–25

**10. ARTICLE IN A WEEKLY JOURNAL**

Witmer LM. 2001 Aug 3. Nostril position in dinosaurs and other vertebrates and its significance for nasal function. Science 292 (5531):850–3.

10. Witmer LM. Nostril position in dinosaurs and other vertebrates and its significance for nasal function. Science 2001 Aug 3; 292 (5531):850–3.

**11. ARTICLE IN A MAGAZINE**

Lopez B. 2001 Oct. The naturalist. Orion:39–43.

11. Lopez B. The naturalist. Orion 2001 Oct:39–43.

**12. ARTICLE IN A NEWSPAPER**

Mishra R. 2001 Dec 1. First patient with full artificial heart dies. Boston Globe; Sect A:2 (col 1).

12. Mishra R. First patient with full artificial heart dies. Boston Globe 2001 Dec 1; Sect A:2 (col 1).

**13. ARTICLE WITH NO IDENTIFIED AUTHOR**

[Anonymous]. 2001 Oct. The bread basket. Nutri Act News:14–5.

13. [Anonymous]. The bread basket. Nutri Act News 2001 Oct: 14–5.

## Electronic sources

The CSE's *Scientific Style and Format* gives only brief coverage of formats for electronic sources, mainly because it was published in 1994. For additional models and formats endorsed by the American Medical Association, the World Association of Medical Editors, and other science-oriented organizations, consult the Vancouver style for electronic citations summarized at <http://www.ama-assn.org/public/peer/wame/uniform.htm>.

**14. SOURCE ON CD-ROM**

MacEachren AM, Kraak MJ, editors. 1997. Exploratory cartographic visualization [CD-ROM]. Oxford: Elsevier.

14. MacEachren AM, Kraak MJ, editors. Exploratory cartographic visualization [CD-ROM]. Oxford: Elsevier; 1997.

#### 15. ONLINE JOURNAL ARTICLE

Castro MA, Vega AS, Mulgura ME. 2001 Nov. Structure and ultrastructure of leaf and calyx glands in *Galphimia brasiliensis*. Am J Bot [serial online]. Available from: http://www.amjbot.org/cgi/content/full/88/11/1935. Accessed 2001 Dec 2.

15. Castro MA, Vega AS, Mulgura ME. Structure and ultrastructure of leaf and calyx glands in *Galphimia brasiliensis.* Am J Bot [serial online] 2001 Nov. Available from: http://www.amjbot.org/cgi/content/ full/88/11/1935. Accessed 2001 Dec 2.

#### 16. ONLINE BOOK OR MONOGRAPH

Margulis L, Schwartz KV, Dolan M. 1999. Diversity of life: the illustrated guide to the five kingdoms [book online]. Boston: Jones and Bartlett. Available from: http://emedia.netlibrary.com. Accessed 2001 Oct 7.

16. Margulis L, Schwartz KV, Dolan M. Diversity of life: the illustrated guide to the five kingdoms [book online]. Boston: Jones and Bartlett; 1999. Available from: http://emedia.netlibrary.com. Accessed 2001 Oct 7.

### Other sources

#### 17. GOVERNMENT DOCUMENT

Committee on Rules, House (US). 2001. Providing for consideration of HR 2505: human cloning prohibition act of 2001. Washington: US GPO. 5 p.

17. Committee on Rules, House (US). Providing for consideration of HR 2505: human cloning prohibition act of 2001. Washington: US GPO; 2001. 5 p.

#### 18. DISSERTATION

Fritzen DE. 1995. Ecology and behavior of mule deer on the Rosebud Coal Mine, Montana [dissertation]. Bozeman (MT): Montana State University. 143 p. Available from: University Microfilms, Ann Arbor, MI: DA 9622310.

18. Fritzen DE. Ecology and behavior of mule deer on the Rosebud Coal Mine, Montana [dissertation]. Bozeman (MT): Montana State University; 1995. 143 p. Available from: University Microfilms, Ann Arbor, MI; DA 9622310.

#### 19. SOUND RECORDING, VIDEO RECORDING, OR FILM

WGBH/NOVA. 2001. Evolution: a journey into where we're from and where we're going [videocassette]. Boston (MA): WGBH Boston Video. 8 videocassettes: 600 min, sound, color, 1/2 in.

19. WGBH/NOVA. Evolution: a journey into where we're from and where we're going [videocassette]. Boston (MA): WGBH Boston Video; 2001. 8 videocassettes: 600 min, sound, color, 1/2 in.

# 56 Other Documentation Style Manuals

Many academic disciplines publish a style manual for their particular system of documentation. You will find these manuals in the reference section of the library.

American Institute of Physics. *AIP Style Manual for Guidance in Writing, Editing, and Preparing Physics Manuscripts for Publication*. 4th ed. New York: Amer. Institute of Physics, 1990. Available online with 1997 addendum at <http://www.aip.org/pubservs/style/4thed/toc.html>.

American Mathematical Society. *AMS Author Handbook*. Rev. ed. Providence: Amer. Mathematical Soc., 1996.

American Political Science Association. *Style Manual for Political Science*. Rev. ed. Washington: Amer. Political Science Assn., 1993.

American Psychological Association. *Publication Manual of the American Psychological Association*. 5th ed. Washington: Amer. Psychological Assn., 2001.

Bates, Robert L., Rex Buchanan, and Marla Adkins-Heljeson, eds. *Geowriting: A Guide to Writing, Editing, and Printing in Earth Science*. 5th ed. Alexandria: American Geological Institute, 1995.

*The Bluebook: A Uniform System of Citation*. Comp. editors of Columbia Law Review, et al. 16th ed. Cambridge: Harvard Law Review Association, 1996.

*The Chicago Manual of Style*. 14th ed. Chicago: U of Chicago P, 1993.

Council of Biology Editors Style Manual Committee. *Scientific Style and Format: The CSE Manual for Authors, Editors, and Publishers*. 6th ed. New York: Cambridge UP, 1994.

Dodd, Janet S., ed. *The ACS Style Guide: A Manual for Authors and Editors*. 2nd ed. Washington: Amer. Chemical Soc., 1986.

Gibaldi, Joseph. *MLA Handbook for Writers of Research Papers*. 5th ed. New York: MLA, 1999.

Goldstein, Norm, ed. *The Associated Press Stylebook and Libel Manual: With Appendixes on Copyright Guidelines, Freedom of Information Act, Photo Captions, Filing the Wire,* 33rd ed. New York: Associated Press, 1998.

Holoman, D. Kern, ed. *Writing About Music: A Style Sheet from the Editors of 19th-Century Music,* Berkeley: U of California P, 1988.

Linguistics Society of America. "LSA Style Sheet." Published yearly in the December issue of the *LSA Bulletin.*

Turabian, Kate L. *A Manual for Writers of Term Papers, Theses, and Dissertations.* 6th ed. Chicago: U of Chicago P, 1996.

# *Glossaries*

# 57 Glossary of Usage

The following glossary of usage is a handy reference guide to pairs of frequently confused words (*disinterested, uninterested*), frequently misspelled words (*their, there, they're*), words that cause subject-verb or pronoun agreement problems (*data, phenomena*), nonstandard words and phrases (*alright, is when*), and informal usages (*kind of, real*). If you have a usage problem not covered here, consult your college dictionary.

**a, an** Use *a* before words that begin with a consonant sound, even if the word begins with a vowel: *a book, a ladder, a unique choice.* Use *an* before a word that begins with a vowel sound, even if the word begins with a consonant: *an idea, an unlikely candidate, an hour.* With a word that begins with a hard *h,* use *a: a hotel;* with a word that begins with an unpronounced *h,* use *an: an honor.*

**accept, except** *Accept* is a verb that means "to receive." *Except* is usually a preposition that means "other than" or "with the exclusion of." *Carla will accept all the invitations except Jason's.*

**advice, advise** *Advice* is a noun and means "opinion about a course of action." *Advise* is a verb and means "to offer advice." *She advised her roommate to seek advice.*

**affect, effect** *Affect* as a verb means "to influence." *Staring at a computer screen can affect your eyesight. Effect* as a noun means "result." *Painting the kitchen yellow brought about the effect she wanted. Effect* as a verb means "to bring about." *Negotiation was the only way to effect the release of the hostages.*

**all ready, already** *All ready* means "prepared." *The swimmers were all ready to start. Already* means "previously or before." *We had already finished dessert when Raphael picked up his salad fork.*

**all right, alright** *All right* is always written as two words, just as *all wrong* is. *Alright* is nonstandard English. *It's all right* [NOT *alright*] *if we meet at the restaurant.*

**allusion, illusion, delusion** An *allusion* is "an implied or indirect reference." *The lawyer made an allusion to a mystery by Agatha Christie.* An *illusion* is "a false concept" or "deceptive impression." *The magician was a master of illusion.* A *delusion* is "a mistaken belief," usually as a result of psychological problems. *He suffered the delusion of thinking he was Napoleon.*

**a lot, alot** *A lot* is two words, not one. *I ate a lot* [NOT *alot*] *of cake.*

**altogether, all together** *Altogether* means "completely." *That's altogether wrong. All together* means "in a group." *All together now, "Row, row, row your boat."*

**among, between** Use *among* for three or more persons or things. *You can choose among several research topics.* Use *between* when there are only two persons or things. *The choice is between coffee and tea.*

**amount, number** *Amount* refers to things in bulk or mass. *A staggering amount of trash filled the basement. Number* refers to things that are countable. *A number of students were given parking tickets.*

**and/or** An awkward and sometimes imprecise example of legal or business usage that is best avoided in college writing. In nonbusiness language, instead of *Mr. Hayes told Marcia that she would receive a promotion and/or raise,* write *Mr. Hayes told Marcia she would receive a promotion or a raise or both.*

**anxious, eager** *Anxious* is often used when *eager* is the appropriate choice. *Anxious* means "filled with worry and apprehension." *Eager* means "looking forward with pleasure." *I am eager* [NOT *anxious*] *to see the Hockney exhibit.*

**anyone, any one** *Anyone* means "any person at all." *Any one* refers to a particular person or thing in a group. *Anyone in the room can select any one of these gifts.*

**anyway, any way** *Anyway* means "nevertheless." *Anyway, who cares? Any way* means "by whatever means." *He'll go any way he can.*

**anyways, anywheres** Nonstandard forms of *anyway* and *anywhere. Ted's going anyway* [NOT *anyways*].

**as, like** See **like, as, as if.**

**awhile, a while** Use the adverb *awhile* to modify a verb. *Rest awhile, if you like.* Use the article and noun *a while* as the object of a preposition. *We went to the park for a while.*

**bad, badly** *Bad* is an adjective, and *badly* is an adverb. *I feel bad that Cathy didn't go.* (If you *felt badly* your sense of touch would be faulty.) *His car was badly damaged.*

**being that, being as (how)** Nonstandard expressions used in place of the subordinating conjunction *because. Because* [NOT *being that*] *Patrick had to go to work early, he couldn't give her a ride.*

**beside, besides** *Beside* is a preposition meaning "next to" or "at the side of." *The first lady sat beside the president. Besides* is an adverb or a preposition meaning "in addition" or "moreover." *Besides free admission, membership includes a discount at the museum gift shop.*

**between, among** See **among, between.**

**bring, take** Use *bring* when something is being moved toward the speaker. *Bring your calculator to class tomorrow.* Use *take* when something is being moved away from the speaker. *Would you please take these bottles to the recycling center?*

**can, may** *Can* means "having ability." *She can play chess. May* indicates "having permission." *The children may watch television.*

**compare to, compare with** *Compare to* means "to represent as similar." *The human brain is often compared to a computer. Compare with* means "to examine the character or quality of two things to see how they are similar or different." *Muhammad Ali's boxing style has been compared with Joe Frazier's.*

**complement, compliment** *Complement* means "to fill out or make whole." *Her lyrics complement his music. Compliment* means "to praise or congratulate." *She always compliments her students when they perform well.*

**conscience, conscious** *Conscience* is a noun meaning "a sense of the difference between right and wrong." *Conscious* is an adjective meaning "awake or alert." *I was conscious that my conscience was being tested.*

**continual, continuous** *Continual* means "recurring at intervals, intermittently." *Max continually tells the story of how we met. Continuous* means "occurring without interruption, incessantly." *She was plagued by a continuous ringing in her ears.*

**could of** Nonstandard for *could have. Craig could have* [NOT *could of*] *gone to summer school.*

**criteria** *Criteria* means "standards or tests on which judgments can be based" and is the plural of *criterion. There is only one criterion for admittance: age. Professor Kim explained the criteria for receiving an* A.

**data** *Data* means "pieces of information" and is the plural of *datum,* which is now rarely used, except in formal writing. Some writers treat *data* as both singular and plural; careful writers treat it as plural only. *The national census data were gathered last year.* For the singular of data, many writers prefer "a piece of data" to *datum.*

**different from, different than** Use *different from* in most instances. *This chili recipe is different from mine.* Use *different than* if a clause follows. *The exam was different than what I expected.*

**disinterested, uninterested** *Disinterested* means "free of self-interest or bias." *She was a disinterested observer at the trial. Uninterested* means "without interest." *I'm uninterested in classical music but love opera.*

**don't** *Don't* is an acceptable contraction for *do not* but not for *does not. You don't want to go, and she doesn't* [NOT *don't*] *want to stay.*

**due to** *Due to* means "owing to" or "because of." It is acceptable as an adjective phrase following some form of the verb *to be: His death was due to natural causes. Due to* is not acceptable as a preposition meaning "because of." *Our flight was grounded because of* [NOT *due to*] *bad weather.*

**eager, anxious** See **anxious, eager.**

**effect, affect** See **affect, effect.**

**e.g.** This is the Latin abbreviation for *exempli gratia,* meaning "for example." It is acceptable in parenthetical comments (*e.g., red, yellow, and blue*), but outside parentheses, *for example* should be spelled out. *Our house, for example, needs a new roof.* (Both *e.g.* and *for example* are always followed by a comma.)

**emigrate from, immigrate to** *Emigrate* means "to leave one country or area for another" and takes the preposition *from. All of my grandpar-*

*ents emigrated from Italy. Immigrate* means "to enter and settle permanently in another country" and takes the preposition *to. They immigrated to America in the 1890s.*

**enthused, enthusiastic** *Enthused* is not widely accepted as an adjective meaning "showing enthusiasm." Use *enthusiastic* instead. *He gets enthusiastic* [NOT *enthused*] *about three things—breakfast, lunch, and dinner.*

**etc.** *Etc.* (*et cetera*) is Latin for "and other things." Do not use *and etc.* because it is redundant. Also, do not use *etc.* to refer to people. Actually, it is stylistically preferable to use the expression *and so on* instead of *etc.* outside parentheses. *Try varying your routine at the gym with the treadmill, stairs, rower, and so on.*

**everyone, every one** See **anyone, any one.**

**except, accept** See **accept, except.**

**farther, further** Use *farther* when distance is involved. *The sailboat drifted farther away.* Use *further* to mean "more" or "to a greater extent." *The neurologist inquired further into Renee's medical history.*

**feel, think** *Feel* means "to be aware of by instinct or inference" and should not be used interchangeably with *think* or *believe. The economist believes* [NOT *feels*] *that inflation is under control.*

**fewer, less** *Fewer* means " a smaller number." *Less* is a comparative of "little." Use *fewer* with items that can be counted, and *less* with amounts. *With fewer* [NOT *less*] *tourists in town, there was less traffic congestion.*

**further** See **farther, further.**

**good, well** *Good* as an adjective means "having beneficial or desirable qualities." *Well* as an adverb means "in a good manner, correctly." *The old Corvette not only looked good, but it ran well.* Use *well* in matters of health. *Do you feel well?*

**hanged, hung** *Hanged* is the past tense and past participle of *hang,* meaning "to execute." *He was convicted and hanged. Hung* is the past tense and past participle of *hang,* meaning "to fasten, attach, or suspend." *Scarlet and gold banners hung from the pulpit.* See pages 60–62.

**he/she, his/her** See page 69 for a full discussion of these pronouns and sexist language.

**hisself, theirselves** *Hisself* and *theirselves* are nonstandard for *himself* and *themselves. Doug can open up the cabin by himself* [NOT *hisself*].

**hopefully** *Hopefully* means "in a hopeful way." Do not use *hopefully* without indicating *who* is being hopeful. *The machinists look hopefully to the end of the strike.* Or, better because it is more direct, *The machinists hope the strike ends soon.* [NOT *Hopefully, the strike will be over soon.*]

**hung** See **hanged, hung.**

**i.e.** This Latin abbreviation for *id est* means "that is." It is used only within parentheses. *Arne Naess popularized "deep ecology"—that is,* [NOT *i.e.,*] *philosophic ecology.* (A comma is usually required after *i.e.* and after *that is.*)

**illusion, allusion** See **allusion, illusion, delusion.**

**immigrate to, emigrate from** See **emigrate from, immigrate to.**

**imply, infer** *Imply* means "to hint" or "to express indirectly." *In her annual address, the president implied she would retire soon. Infer* means "to conclude from evidence." *From what the financial officer said, students inferred it would be more difficult to get loans.*

**incredible, incredulous** *Incredible* means "unbelievable." *Incredulous* means "disbelieving." *The judge was incredulous that the defense attorney was so incredibly unprepared.*

**infer, imply** See **imply, infer.**

**in regards to** A mixture of *in regard to* and *as regards.* Use one phrase or the other. Both, however, sound very formal, if not inflated; use *about* instead. *Talk to your commanding officer about* [NOT *in regards to*] *reenlisting.*

**irregardless** Nonstandard for *regardless. Regardless* [NOT *irregardless*], *wear neon if you run at night.*

**is when, is where** Do not use *when* and *where* after *is* in definitions. *Scuttling a ship is a process by which* [NOT *is when*] *holes are cut in its bottom to sink it.*

**its, it's** *Its* is the possessive form of *it. It's* is the contraction of *it is.* Test whether you are using the correct spelling by inserting the words *it is. The Alaskan wolf is losing more of its* [NOT *it's*] *habitat every year.*

**kind of, sort of, type of** Used colloquially, these expressions mean "rather" or "somewhat." They are acceptable in speech but can add unnecessary words in writing. *I was somewhat* [NOT *kind of*] *disappointed by the low attendance.* Or, simply, *I was disappointed by the low attendance.* When you use these expressions to classify, do not use an *a. That kind of* [NOT *kind of a*] *potato is best for baking.*

**lay, lie** *Lay* means "to put" or "to place" and always takes an object (principal parts: *lay, laid,* and *laid*). *Lie* means "to recline" and is an intransitive verb, meaning it does not take an object (principal parts: *lie, lay,* and *lain*). Usage errors occur because the present tense of *lay* and the past tense of *lie* are both *lay. Lie* [NOT *lay*] *down immediately. The minute she lay* [NOT *laid*] *down, she fell asleep. Lay the carpet carefully.*

**less, fewer** See **fewer, less.**

**lead, led** *Lead* is a noun meaning "a type of metal." *Led* is the past tense of the verb *to lead. The crossing guard led* [NOT *lead*] *us through the intersection.*

**like, as, as if** *Like* is a preposition expressing similarity. *She looks like Julia Roberts. As* and *as if* are subordinating conjunctions and are used to introduce dependent clauses. *As* [NOT *like*] *he said, he was out of town at the time.*

**loose, lose** *Loose* is an adjective meaning "not tight." *Lose* is a verb and means "to be unable to find" or "to be defeated in a game." Confusion comes about as much from misspelling as from misuse. *Did you lose the top button? It looked loose.*

**may, can** See **can, may.**
**may of** Nonstandard for *may have.* See **could of.**
**maybe, may be** *Maybe* is an adverb meaning "possibly or perhaps," and *may be* is a verb. *Wednesday's field trip may be a good one; maybe I'll go.*
**media, medium** *Media* is the plural of *medium,* meaning "a publication or broadcast." *Television is the medium that most influences the American people.*
**might of** Nonstandard for *might have.* See **could of.**
**myself, himself/herself, yourself** These are reflexive pronouns (*I saw myself reflected in the store window*) or intensive pronouns (*I myself would let bygones be bygones*). Do not, however, use them as substitutes for *me, he/she,* or *you. Packages arrived for Maria and me* [NOT *myself*].
**nowheres** Nonstandard for *nowhere.* See **anyways, anywheres.**
**number, amount** See **amount, number.**
**off of, off from** The *of* and *from* are redundant. *The cans fell off* [NOT *off of* OR *off from*] *the shelf.*
**OK, okay** In formal speech and writing, use a word such as *approval* or *agreement* instead. *Our zoning board approved* [NOT *okayed*] *the new subdivision.*
**phenomena** *Phenomena* means "observable facts or events" and is the plural of *phenomenon.* There is no such word as *phenomenas. Edgar Allan Poe's short stories are filled with strange phenomena.*
**plus** *Plus* means "added to": *The bill was $90 plus tax.* Do not use *plus* to link independent clauses. *Lethal injection is cruel and unusual punishment, and* [NOT *plus*] *it does not deter murderers.*
**precede, proceed** *Precede* means to "come or go before," and proceed means "to advance or move forward." *As the baseball team proceeded down the street, they could hear the band that preceded them.*
**principal, principle** As an adjective, *principal* means "first or foremost, chief." *Myron is the principal cellist.* As a noun, *principal* means "leader" or "a sum of money as distinguished from interest or profit." *Mr. Marcotte is the new principal. The idea is to live off the interest and not the principal. Principle* is always a noun and means "a truth, or basic assumption." *She lives by the principle that less is more.*
**quote, quotation** *Quote* is a verb; *quotation* a noun. In formal writing, do not use *quote* as a shortened form of *quotation. His term paper was a string of quotations* [NOT *quotes*].
**real, really** In speech, *real* is used for "very" or "extremely." In formal writing, use *really. Jay Leno is really* [NOT *real*] *funny.* Avoid overusing the adverb.
**reason is because, reason why** These expressions are awkward and redundant. Simply omit them. *I'm not going away for my vacation because I don't have the money.* [NOT *The reason why I'm not going away for my vacation is because I don't have the money.*]

**sensual, sensuous** *Sensual* pertains to bodily or sexual pleasure. *Balanchine's choreography is very sensual. Sensuous* means "appealing to the senses." *The smell of bread baking is positively sensuous.*

**set, sit** *Set* means "to put something down" and takes a direct object (principal parts: *set, set, set*); *sit* means "to rest or be seated" (principal parts: *sit, sat, sat*). *Sit here, and set your popcorn on that little table.* See page 63.

**should of** Nonstandard for *should have.* See **could of.**

**since** *Since* means "between then and now" and should not be used to mean *because. Because* [NOT *since*] *the plane needed repairs, the tourists had not been able to leave.*

**sit, set** See **set, sit.**

**someone, some one** See **anyone, any one.**

**somewheres** Nonstandard for *somewhere.* See **anyways, anywheres.**

**sort of, kind of** See **kind of, sort of, type of.**

**sure and** Nonstandard. Use *sure to* instead. *Be sure to* [NOT *sure and*] *proofread business letters carefully.*

**take, bring** See **bring, take.**

**than, then** *Than* is a conjunction used in a clause of comparison or inequality. *I would rather have ice cream than* [NOT *then*] *cake. Then* is an adverb used to indicate past or future time. *We'll do our errands and then have lunch.*

**that, which** *That* is used to introduce restrictive clauses. *The cowboy poetry that I heard last night was very funny. Which* is used to introduce nonrestrictive clauses. *His morals, which are questionable, should not influence our opinion of this novel.* See page 58.

**their, there, they're** *Their* is a possessive pronoun. *Their sidewalk is always shoveled in winter. There* means "at or in that place." *We'll call when we get there. There* is also an expletive (*there is/are . . .* ). *There is reason to believe Gwen is in Mexico. They're* is a contraction for "they are." *They're at their house now.*

**theirself, theirselves** See **hisself, theirselves.**

**to, too, two** Perhaps more of a spelling than a usage problem. *To* is a preposition, *too* is an adverb, and *two* is a number. *Two tickets to a Metallica concert are two too many for my parents.*

**toward, towards** Both versions are acceptable. *Toward* is preferred in the United States. *We rowed steadily toward shore.*

**try and** Nonstandard. Use *try to* instead. *Try to* [NOT *Try and*] *be on time.*

**uninterested, disinterested** See **disinterested, uninterested.**

**unique** Like *perfect* and *complete, unique* is an absolute. There are no degrees of uniqueness; either something is one of a kind, or it isn't. *Very unique* is not only illogical but an overused intensifier. *Dilbert cartoons are unique* [NOT *very unique*] *in their ridiculous humor.*

**use to, suppose to** Nonstandard for *used to* and *supposed to. Jessica used* [NOT *use*] *to date my brother.*

**wait on** *Wait on* means "to serve" and should not be used when the desired meaning is "wait for." *We waited for* [NOT *on*] *Liz, but she never showed up.*

**ways** Colloquial; use *way* when designating a distance. *You have a way* [NOT *ways*] *to go before you get to I-85.*

**which, that** See **that, which.**

**which, who, whose, that** Generally, use *who* or *whose* to refer to people, except in an expression such as *That's an idea whose time has come.* Use *which* or *that* to refer to things. Do not use *which* to refer to people. *It was the tall man with red hair who* [NOT *which*] *witnessed the robbery.*

**who's, whose** *Who's* is a contraction of *who is. Whose* is a possessive pronoun. *Whose* [NOT *who's*] *volleyball is this?*

**would of** Nonstandard for *would have.* See **could of.**

**you** In formal writing, avoid the use of *you* to mean "anyone." *In most civic organizations, each volunteer is* [NOT *you are*] *expected to do some kind of charitable work.*

**your, you're** *Your* is a possessive pronoun. *You're* is a contraction of *you are. You're* [NOT *your*] *scheduled for the 6:00 flight.*

# 58 Glossary of Grammatical Terms

A boldface term within a definition indicates that the term has its own definition within this glossary.

**absolute phrase** A noun or noun equivalent that is followed in most cases by a **participial phrase.** An absolute phrase modifies a **clause** or **sentence,** not just a word as all other phrases do. *Her career progress grinding to a halt, Juanita decided to take some time off.*

**active voice versus passive voice** In the active voice, the **subject** of the sentence does the acting; in the passive voice, the subject is acted upon. Active: *The doctor examined the X rays.* Passive: *The X rays were examined by the doctor.*

**adjective** A word that modifies or qualifies a **noun** or **pronoun.** An adjective tells what kind, how many, or which one: *yellow pad* [what kind], *101 Dalmatians,* [how many], *the biggest one* [which one].

**adjective clause** An adjective clause (also called a relative clause) modifies a **noun** or **pronoun.** Usually it immediately follows the word or words it modifies and begins with a **relative pronoun** (*who, whose, whom, which, that, whoever, whomever,* or *whatever*) or with the **subordinating conjunction** *when* or *where. Erik reminded me of Holden Caulfield, who also wore his baseball hat backward.*

**adverb** A word that modifies a **verb,** an **adjective,** or another adverb. Adverbs tell where, when, how, why, under what circumstances, and to what extent: *carefully paints* [adverb modifying a verb], *especially clever* [adverb modifying an adjective], *very calmly* [adverb modifying an adverb].

**adverb clause** An adverb clause modifies a **verb,** an **adjective,** or an **adverb.** It always begins with a **subordinating conjunction** (e.g., *if, after, when, though, since, where,* or *while*). An adverb clause tells when, where, why, under what circumstances, or to what extent: *The firemen left after the fire was extinguished.*

**agreement** See **pronoun–antecedent agreement; subject–verb agreement.**

**antecedent** The **noun** or **pronoun** to which a pronoun refers. *Liz, who works for a consulting firm, seldom knows where she will work next. Liz* is the antecedent of the pronouns *who* and *she*.

**appositive** A noun or pronoun equivalent that directly follows, or sometimes precedes, another noun or noun equivalent and serves to amplify its meaning. *Allie, the captain of the women's hockey team, is an excellent skater.*

**article** The words *a, an* (indefinite articles), and *the* (definite article) used to mark a noun. Articles are considered **adjectives.**

**auxiliary verb** See **helping verb.**

**case** See **objective case; possessive case; subjective case.**

**clause** A group of words that contains a **subject** and a **predicate.** A **main clause,** or independent clause, can stand alone as a complete **sentence.** *Jake bought a used car.* A **subordinate clause,** or dependent clause, cannot stand alone as a sentence because it is an incomplete thought. *When I shop, I avoid malls.*

**collective noun** A word, singular in form, that names a group of individuals or things: *army, band, class, family.* If a collective noun refers to the group as a whole, treat it as singular. If it refers to the multiple members of the group, treat it as plural. Singular: *The committee gives its report today.* Plural: *The committee give themselves annual raises.*

**comma splice** A sentence fault in which only a comma or a comma and a **conjunctive adverb** are used to join two or more **main clauses.** Comma splice: *Sharonna got a good grade, it was a difficult assignment.* Revised: *Sharonna got a good grade; it was a difficult assignment. Sharonna got a good grade, and it was a difficult assignment.*

**comparative forms of adjectives and adverbs** See pages 77–78.

**complement** See **object complement; subject complement.**

**complex sentence** See **sentence.**

**compound-complex sentence** See **sentence.**

**compound sentence** See **sentence.**

**conjunction** A word used to join words or groups of words in a **sentence.** There are three types of conjunctions—**coordinating, subordinating,** and correlative—and **conjunctive adverbs.**

**conjunctive adverb** An adverb used with a semicolon to link two **main clauses** in one sentence: *Lisa thought she'd pick up a pizza for dinner; meanwhile, Todd was home cooking her favorite—fried chicken.* Common conjunctive adverbs: *accordingly, also, finally, however, nevertheless, thus.*

**coordinating conjunction** A conjunction that connects words, phrases, and clauses of equal rank: *and, for, or, yet, but, nor, so.*

**correlative conjunction** Conjunctions that work in pairs to connect grammatically equivalent elements: *either . . . or, neither . . . nor, not only . . . but also.*

**count noun** A **noun** that names items that can be counted: *books, cars, sentences.* Frequently, count nouns are accompanied by an **adjective** indicating how many: *four books.*

**dangling modifier** A **phrase** that does not logically relate to the main part of the sentence. Dangling: *After sitting down, the exam began.* Revised: *After sitting down, the students began the exam.*

**demonstrative pronoun** The **pronoun**—*this, that, these,* or *those*—that points out the **noun** it replaces in a **sentence** such as *That hurts!*

**dependent clause** See **subordinate clause.**

**direct address** A construction in which a person spoken to is mentioned. *"Nice basket, Caroline."*

**direct object** The word or word group that receives the action of the verb. *He kissed her.*

**double negative** A nonstandard construction using two negatives. *He didn't say nothing.*

**ellipsis mark** Three spaced periods that indicate a word or words have been omitted from a quotation. *"Give me liberty or . . . death," said Patrick Henry.*

**expletive** The word *there, here,* or *it* used in the beginning of a **sentence** to delay the **subject:** *There is a large Hmong community in Minneapolis.*

**fused sentence (run-on sentence)** A sentence fault that occurs when no punctuation and no **coordinating conjunction** appear between two or more **main clauses.** Fused: *The legislators could not agree on the bill they had to stay in session.* Revised: *The legislators could not agree on the bill. They had to stay in session.*

**future progressive** See **tense.**

**future perfect tense** See **tense.**

**future tense** See **tense.**

**gerund** A **verb** form that ends in *-ing* and that functions as a **noun.** *Sailing can be a very relaxing sport.*

**gerund phrase** A **gerund** (e.g., *brewing, flying, joking, studying*) plus any **modifier**(s) and/or object(s): *All this fancy cooking takes time. Cooking a goose is messy work.*

**helping verb (auxiliary verb)** A verb used with a **main verb.** The most common helping verbs are the forms of *be, have,* and *do.* Other helping verbs include the nine **modals:** *may, might, shall, will, would, must, should, can, could.* Helping verbs always appear before the main verb: *will eat, is eating, had eaten.*

**idiom(atic)** An expression that has a meaning different from the literal meanings of its individual words: *to catch a cold; to give someone a piece of your mind; it's raining cats and dogs.*

**imperative mood** See **mood.**

**indefinite pronoun** A **pronoun** that refers to a nonspecific person or thing (e.g., *all, anybody, both, everybody, nothing, several, some. Everybody was upset about losing the game*).

**independent clause** See **main clause.**

**indicative mood** See **mood.**

**indirect object** The **noun, pronoun,** or **verbal** that tells *to whom* or *for whom* something is done. *The coach gave Melissa another chance.*

**infinitive** The base form of a **verb** preceded by *to: to farm, to swim, to think.*

**infinitive phrase** An **infinitive** along with its **objects, complements,** and **modifiers.** An infinitive phrase can function as a **noun,** an **adjective,** or an **adverb:** *To think compassionately is to live a good life.*

**interjection** A word or phrase used to express emotion or to attract attention: *Hey! There's a parade coming.*

**interrogative pronoun** A **pronoun** used to introduce a question: *who(ever), whom(ever), whose, which(ever), what(ever).*

**intransitive verb** An intransitive verb does not require an object or complement to complete its meaning: *Nobody cares about me.*

**irregular verb** See **regular and irregular verbs.**

**linking verb** A **verb** that connects the **subject** of a **sentence** to a predicate noun, which renames the subject, or to a predicate adjective, which describes the subject. *Jamaica Kincaid is a writer. The grapefruit tastes sour.*

**main clause (independent clause)** A group of words that can stand alone as a complete **sentence.** *Janie bought Ted's old Volkswagen.*

**main verb** The part of the verb phrase that expresses the primary meaning: *has been sleeping, might change, was thrilled.* See **helping verb.**

**misplaced modifier** A **modifier** that is not as close to the word or words it modifies as it should be. Misplaced: *The waiter brought a steak to the young man covered with mushrooms.* Revised: *The waiter brought a steak covered with mushrooms to the young man.*

**modal** A modal expresses how the writer feels about an action. A modal expresses probability, necessity or obligation, or ability. See **helping verb.**

**modifier** A word, **phrase,** or **clause** that describes a **noun, verb, phrase,** or **clause.** Modifiers include **adjectives** and **adverbs, prepositional phrases, participial phrases, adjective clauses,** and **adverb clauses.**

**mood** The three moods of a verb—indicative, imperative, and subjunctive—reveal the writer's intent, that is, how the writer views a thought or action. The indicative mood makes a statement or asks a question. *It is raining. Is it raining?* The imperative mood gives an order. *Gargle with salt water.* The subjunctive mood expresses a demand, a wish, a condition contrary to fact. *If I were younger, I would go to law school.*

**noncount noun** A **noun** that names items that cannot be counted without the use of another noun: *butter, sugar, anger, stress;* to count such nouns, English requires a construction such as *two pats of butter; two outbursts of anger.*

**nonrestrictive element** A nonrestrictive **phrase, clause,** or **appositive** does not contain essential information that limits—or restricts—the meaning of the **noun** or **pronoun** it describes. A nonrestrictive element is usually set off by commas whether it occurs in the beginning, middle, or end of a sentence: *Awake since dawn, the nervous groom tossed and turned.*

**noun** The name of a person, place, thing, quality, action, or idea: *George, Brazil, chair, beauty, flight, mercy.* See **count noun; noncount noun; proper noun.**

**noun clause** A group of words that has a **subject** and **predicate** and that can function in a way that a **noun** does—as a subject, a **subject complement,** an **indirect object,** an **object complement,** an **object of a preposition,** or an **appositive.**

**number** Number refers to whether a word is singular or plural and is indicated by the form of that word. Singular: *I, she, that, mouse, sings.* Plural: *we, they, those, mice, sing.*

**object** See **direct object; indirect object; object of a preposition.**

**object complement** A word or word group that renames or modifies a **direct object,** always appearing after the direct object. *The judge declared the dolls collectibles. Heraldo painted his car green.*

**objective case** The objective case, or form, of a **personal pronoun** can function as a **direct object** (*Mary invited him*), **indirect object** (*Julie sent him flowers*), or **object of a preposition** (*between Betsy and me*).

**object of a preposition** An object of a preposition is a **noun** or **pronoun** that follows a **preposition.** *Juanita went to the store.*

**parallelism** The use of similar grammatical forms to reinforce or emphasize two or more coordinated words, **phrases,** or **clauses.** *Early to bed and early to rise makes a person healthy, wealthy, and wise.*

**participial phrase** A **present participle** or **past participle** with its **auxiliary, modifier**(s), **object**(s), and **complement**(s). A participial phrase always functions as an **adjective.** *The woman riding the mountain bike is my cousin. The woman blocked by the horse is my cousin.*

**participle** See **past participle; present participle.**

**parts of speech** The categories into which we classify words according to their form, function, and meaning in sentences: **verbs, nouns, pronouns, adjectives, adverbs, prepositions, conjunctions,** and **interjections.**

**passive voice** See **active voice versus passive voice.**

**past participle** The past participle is the *-ed* form of regular verbs: *cramped, traveled, celebrated.* Irregular verbs form their past participles variously: *swum, chosen, grown, paid.*

**past perfect tense** See **tense.**

**past progressive** See **progressive tenses.**

**past tense** See **tense.**

**perfect tense** See **tense.**

**person** First-person pronouns (*I, we, me, us, my, our*) highlight the writer and are appropriate for experiential writing. Second-person pronouns (*you, your, yours*) direct attention to the reader and are appropriate for directions, instructions, and other how-to types of writing. Third-person pronouns (*he, she, it, they, him, her, them, his, hers, its, their, theirs*) highlight the **subject** and are appropriate for most informative-writing situations, including academic writing.

**personal pronoun** A **pronoun** that substitutes for a specific **noun** or other pronoun, thus referring to specific persons, places, or things: *I, me, you, he, she, it, we, us, you, they, them.*

**phrase** The phrase, the most common word group in English writing, lacks a **subject,** a **verb,** or both. Phrases function within **sentences** as **adjectives,** as **adverbs,** and as **nouns.** See **absolute phrase; appositive; gerund phrase; infinitive phrase; participial phrase; prepositional phrase; verbal phrase.**

**possessive case** The possessive case, or form, of **nouns** and **pronouns** indicates ownership: *the children's pet, her salary.*

**possessive pronoun** Possessive pronouns are personal pronouns that show ownership: *my, mine, your, yours, his, her, hers, its, our, ours, your, yours, their, theirs.*

**predicate** A predicate is the **sentence** part that consists of a **verb** and any **objects, complements,** and **modifiers** that accompany it. *The voters sent the legislature a clear message.*

**preposition** A word that comes before a **noun** or **pronoun** to create a **phrase** that modifies another word in the sentence. Prepositions show relationships between objects and ideas in a sentence. Common prepositions include *about, across, after, among, before, between, during, of, over, since, through, toward, under, upon,* and *with.*

**prepositional phrase** A **preposition** (a word such as *to, with, after,* or *between*) and its object, a **noun** or noun equivalent. A prepositional phrase usually functions as an **adjective** (*The chapter on whaling is richly illustrated*) or as an **adverb** (*Many Floridians travel to Canada in the summer*).

**present participle** The present participle is the *-ing* form of a **verb** used as an **adjective.** *My friend Jake is annoying.*

**present perfect tense** See **tense.**

**present progressive** See **progressive tenses.**

**present tense** See **tense.**

**principal parts of verbs** The principal parts of a verb are the three forms from which we create the various tenses: base form, or **infinitive** (*[to] talk/write*), past tense form (*talked/wrote*), and **past participle** (*talked/written*).

**progressive tenses** The progressive tenses of a **verb** describe continuing action and use the *-ing* form of the verb: *is emptying, was visiting, will be looking.*

**pronoun** A word that takes the place of a **noun** in a **sentence:** *I, she, somebody, who, themselves.* See **demonstrative pronoun; indefinite pronoun; interrogative pronoun; personal pronoun; possessive pronoun; relative pronoun.**

**pronoun–antecedent agreement** A **pronoun** must agree with its **antecedent**—the noun(s) to which it refers—in gender, **number,** and **person:** *Saro and Jon were exhausted after their chemistry final.*

**proper noun** A **noun** that names a specific person, place, or thing. A proper noun requires a capital letter. *Meryl Streep, South Africa, Statue of Liberty.*

**regular and irregular verbs** A regular verb is one that forms both the past tense and the **past participle** by adding *-ed* or *-d* to the base form of the word: *kiss, kissed, kissed.* An irregular verb has different forms for the past tense and past participle: *begin, began, begun; choose, chose, chosen; eat, ate, eaten; lie, lay, lain; see, saw, seen.*

**relative clause** See **adjective clause.**

**relative pronoun** A relative pronoun—*who, whoever, whom, whomever, whatever, whose, which, that*—joins a **subordinate clause** (or dependent clause) to a **noun** or another **pronoun.** *The dog that lives across the street has a loud bark.*

**restrictive element** A restrictive **phrase, clause,** or **appositive** contains essential information that limits or specifies the meaning of the **noun** or **pronoun** it describes. Restrictive elements are not set off by commas. *The book with the beautiful pictures of the Chinese countryside got us excited about our trip.*

**run-on sentence** See **fused sentence.**

**sentence** A sentence is a complete thought consisting of at least a **subject** and a **predicate.** Sentences are commonly classified in terms of their grammatical structure. A simple sentence consists of one **main clause** and no **subordinate clauses.** *Reading and writing are the primary goals of early education.* A compound sentence consists of two or more main clauses and no subordinate clauses. *The ducks were here all summer, but now they have migrated south.* A complex sentence consists of one main clause and one or more subordinate clauses. *If we don't become stewards of our environment, our lives will change.* A compound-complex sentence consists of two or more main clauses and at least one subordinate clause. *Even though weather conditions were not ideal, the fans came out in record numbers, and our team won the annual Thanksgiving game.*

**sentence fragment** A sentence fragment is a group of words presented as if it were a complete **sentence** even though it starts with a subordinating word or lacks a **subject** or **predicate** or both. Sentence followed by sentence fragment: *I wanted to buy both Celene Dion albums. But had enough money to purchase only one.* Revised: *I wanted to buy both Celene Dion albums but had enough money to purchase only one.*

**series** A series is three or more **parallel** words, phrases, or clauses with the same function. Commas are used to separate items in a series. *Willa Cather wrote poetry, fiction, and nonfiction.*

**simple sentence** See **sentence.**

**split infinitive** An **infinitive** is said to be split when one or more **modifiers** are inserted between a base form of a **verb** and its marker *to*. Split: *The president instructed us to very carefully analyze the report.* Revised: *The president instructed us to analyze the report very carefully.*

**squinting modifier** A squinting modifier looks in two directions; that is, it appears to modify both the word it follows and the word it precedes. *The passenger who was hurt badly needed help.* Revised: *The badly hurt passenger needed help.*

**subject** The simple subject of a **sentence** is the **noun** or **pronoun** that names the person, event, or thing about which an assertion is made in the **predicate** of the sentence. *Bill coached the high school track team.* The complete subject is the simple subject and all its **modifiers:** *My son Bill coached the high school track team.*

**subject complement** A word or word group that comes after a **linking verb** and either renames or describes the **subject** of the sentence. *Larry's team is the Yankees* (**noun** as subject complement). *The waiting room seemed sterile* (**adjective** as subject complement).

**subjective case** In the subjective case, a **pronoun** functions as the **subject** (*She received the prize*) or as the **subject complement** (*It was he*) of a **sentence.**

**subject–verb agreement** Subject–verb agreement refers to the correspondence in form between a **verb** and its **subject.** Every verb in a **clause** or **sentence** must agree in **number** and **person** with its subject. *Overcrowding causes many of the discipline problems in our schools. Hurricanes destroy millions of dollars worth of property each year.*

**subjunctive mood** See **mood.**

**subordinate clause (dependent clause)** A group of words that begins with a **subordinating conjunction** (e.g., *because, if, although*) or a **relative pronoun** (e.g., *who, which, that*) and that contains a **subject** and a **verb.** A subordinate clause cannot stand alone as a complete **sentence.** Within sentences, subordinate clauses function as **adjectives, adverbs,** or **nouns.** See **adjective clause; adverb clause; noun clause.**

**subordinating conjunction** A **conjunction** that introduces a **subordinate clause** and indicates its relation to the **main clause.** *Unless we hurry, we won't arrive before the lecture starts.* Common subordinating conjunctions: *after, although, before, in order that, since, though, until, when, while.*

**subordination** Subordination puts the less important ideas in dependent **phrases** or **clauses,** thus emphasizing the important idea in the **main clause.** *When she was thirty years old, Paula made her first solo flight across the Atlantic.* [The less important thought, her age, is subordinate to the main point, what Paula achieved.]

**tense** The tense of a verb indicates the time of an action. The present tense expresses an action occurring in the present, a habitual action, or a fact or general truth. *Sarah works at Georgia State University. Bill runs several miles every morning. Water boils at 212° F.* The past tense indicates an action that occurred entirely in the past. *Betty flew to her daughter's house for the holidays.* The future tense expresses an action that will occur in the future. *The Hillgroves will take their vacation in January this year.* The present perfect tense indicates an action that occurred in the past but continues into or affects the present, or an action that occurred at no specific time in the past. *She has taught master classes in violin for over ten years. Several small recording studios have produced big hits.* The past perfect tense describes an action that preceded another action when both occurred in the past. *Everyone had eaten before I arrived.* And the future perfect tense expresses an action that will be completed before another future event. *Ashley will have accumulated seventy credits in engineering by the end of the semester.* See **progressive tenses.**

**transitional expression** A word or phrase used to link sentences or paragraphs and to show the relationship between them: *finally, for example, afterward, consequently, in summary, likewise.*

**transitive verb** A transitive verb transfers its action from the **subject** to the object of the **sentence,** and it therefore always takes a **direct object** and sometimes an **indirect object** to complete its meaning. *Everybody likes Fran.*

**verb** A word that expresses action (e.g., *run, think, write*) or a state of being (e.g., *is, become, seem*). A verb is the one word necessary in a **predicate** for a **sentence** to be complete: *Mark McGwire slammed seventy home runs.* See **active versus passive voice; mood; number; person; regular and irregular verbs; tense.**

**verbal** There are three types of verbals. See **gerund; infinitive; participle.**

**verbal phrase** There are three types of verbal phrases. See **gerund phrase; infinitive phrase; participial phrase.**

# Index

# Index

For definitions and examples of grammatical terms, consult the *Glossary of Grammatical Terms,* pages 249–257. For questions of usage (e.g., Do I use *affect* or *effect*?), consult the *Glossary of Usage,* pages 242–249. For questions of documentation format, consult the appropriate style directories (MLA, pages 179, 184–185; APA, pages 203–204, 207–208; and Chicago, 220–221).

# Acknowledgments

**Page 113:** From "The Road Not Taken" in *The Poetry of Robert Frost,* edited by Edward Connery Lathem (published by Jonathan Cape), Copyright 1944 by Robert Frost. Copyright 1916, © 1969 by Henry Holt and Company, Inc. and the Estate of Robert Frost.

**Page 114:** Reprinted by permission of the publishers and the Trustees of Amherst College from *The Poems of Emily Dickinson,* Thomas H. Johnson, ed., Cambridge, Mass.: The Belknap Press of Harvard University Press, Copyright © 1951, 1955, 1979 by the President and Fellows of Harvard College.

**Page 119:** The lines from "what if a much of a which of a wind," Copyright 1944, © 1972, 1991 by the Trustees for the E. E. Cummings Trust, from *Complete Poems: 1904–1962* by E. E. Cummings, edited by George J. Firmage. Used by permission of Liveright Publishing Corporation.

**Page 141:** Microsoft Live Clip Art Gallery. "Dove Clip Art." Reprinted by permission of the Microsoft Corporation.

**Pages 156–157:** Google Web Directory. Screenshots of subject directory search on *http://google.com* reprinted by permission of Google, Mountain View, CA.

**Pages 200–202:** Patricia LaRose, "Recycling Used Newsprint: Coming to Terms with an Environmental Problem." Reprinted with permission.

**Pages 217–219:** Brooke Bailey, "Public Opinion and the Death Penalty." Reprinted with permission.

**Pages 230–232:** Jake Jamieson, "The English-Only Movement: Can Americans Proscribe Language with a Clear Conscience?" Reprinted with permission.

# *Revision Checklist*

## Topic/Thesis

- ❒ Is my topic well focused?
- ❒ Does my thesis statement identify my topic and make an assertion about it?
- ❒ Is my thesis statement positioned so that readers will not miss it?

## Purpose/Organization

- ❒ Is my purpose clear? Is my organizational pattern suited to my purpose?
- ❒ Is my organizational pattern easy for my readers to follow?
- ❒ Is the logic of my argument sound?

## Paragraphs

- ❒ Are my paragraphs unified, well-developed, and coherent?
- ❒ Do my topic sentences relate to my thesis? Do they support my thesis?

## Sentences/Diction

- ❒ Do my sentences convey my thoughts clearly?
- ❒ Do my sentences emphasize the most important aspects of each thought?
- ❒ Is the structure of my sentences varied?
- ❒ Do I engage my reader with concrete nouns and strong action verbs?
- ❒ Do I use appropriate language given my subject and my audience?

## Documentation of Sources

- ❒ Is borrowed material introduced by a signal phrase and documented with a parenthetical in-text citation?
- ❒ Have I properly documented all material that is not common knowledge?
- ❒ Are my direct quotations accurate and enclosed within quotation marks or set off in block format?
- ❒ Is my paper free of plagiarism? Have I used my own words and sentence structure for summaries and paraphrases?

# Correction Symbols

*References are to page numbers.*

| | |
|---|---|
| ab | faulty abbreviation, 120–121 |
| ad | improper use of adjective or adverb, 75–78 |
| agr | faulty agreement, 56–60, 68–72 |
| appr | inappropriate word, 37–40 |
| awk | awkward |
| cap | capital letter, 116–120 |
| case | error in case, 72–75 |
| cit | error in citation, 178–239 |
| cl | replace cliché, 37 |
| coh | coherence, 18–20 |
| coord | faulty coordination, 33–35 |
| cs | comma splice, 80–81 |
| det | add details, 14–20 |
| diction | inappropriate diction, 37–40 |
| dev | inadequate development, 16–18 |
| dgl | dangling construction, 33 |
| exact | inexact word/language, 36–37 |
| frag | sentence fragment, 78–79 |
| fs | fused sentence, 80–81 |
| ESL | ESL basics, 81–90 |
| hyph | error in hyphenation, 123–127 |
| ital | italics (underlining), 123–125 |
| jar | jargon, 37–40 |
| lc | use lowercase letter, 116–120 |
| log | error in logic |
| mixed | mixed construction |
| mm | misplaced modifier, 32–33 |
| mood | error in mood, 67 |
| ms | error in manuscript form, 136–150, 196–199, 213–216, 227–229 |
| no ¶ | no new paragraph, 14–20 |
| num | error in use of numbers, 122–123 |
| P | error in punctuation, 92–114 |
| . ? ! | period, question mark, exclamation point, 107–109 |
| ^, | insert comma, 92–97 |
| no ^, | no comma, 92–97 |
| ; | semicolon, 98–100 |
| : | colon, 100–102 |
| ’ | insert apostrophe, 102–104 |
| “ ” | quotation marks, 104–107 |
| - - | dash, 109–110 |
| ( ) | parentheses, 110–111 |
| [ ] | brackets, 111 |
| . . . | ellipsis, 112–113 |
| / | slash, 113–114 |
| par, ¶ | new paragraph, 14–20 |
| pass | ineffective passive voice, 25, 67–68 |
| ref | error in pronoun reference, 70–72 |
| rep | unnecessary repetition, 26 |
| run on | run-on sentence, 80–81 |
| sexist | sexist language, 39–40, 69 |
| shift | distracting shifts, 29–31 |
| sl | slang, 38 |
| sp | misspelling, 128–133 |
| sub | error in subordination, 33–35 |
| t | error in verb tense, 63–67 |
| trans | add transition, 19 |

*(continued)*

# Correction Symbols *(continued)*

*References are to page numbers.*

| | | | |
|---|---|---|---|
| usage | usage problem, 242–249 | ⌇ | delete |
| vag | use specific words, 36 | ∾ | transpose letters or words (teh) |
| vb | error in verb form, 60–63 | ∧ | insert |
| wdy | wordy, 24–27 | // | faulty parallelism, 28–29 |
| ww | wrong word, 36–40 | # | insert space |
| ⊂ | close up space | | |

# Contents

## Writing with Purpose



## Writing with Clarity

## *Punctuation*